PEARSON EDEXCEL INTERNATIONAL A LEVEL
MECHANICS 3
Student Book

Series Editors: Joe Skrakowski and Harry Smith

Authors: Dave Berry, Keith Gallick, Susan Hooker, Michael Jennings, Jean Littlewood, Lee McKelvey, Bronwen Moran, Su Nicholson, Laurence Pateman, Keith Pledger, Joe Skrakowski, Harry Smith, Jack Williams

Published by Pearson Education Limited, 80 Strand, London, WC2R 0RL.

www.pearsonglobalschools.com

Copies of official specifications for all Pearson qualifications may be found on the website: https://qualifications.pearson.com

Text © Pearson Education Limited 2019
Edited by Linnet Bruce
Typeset by Tech-Set Ltd, Gateshead, UK
Original illustrations © Pearson Education Limited 2019
Illustrated by © Tech-Set Ltd, Gateshead, UK
Cover design by © Pearson Education Limited 2019

The rights of Dave Berry, Keith Gallick, Susan Hooker, Michael Jennings, Jean Littlewood, Lee McKelvey, Bronwen Moran, Su Nicholson, Laurence Pateman, Keith Pledger, Joe Skrakowski, Harry Smith, Jack Williams to be identified as the authors of this work have been asserted by them in accordance with the Copyright, Designs and Patents Act 1988.

First published 2019

25 24 23 22 21 20
10 9 8 7 6 5 4 3

British Library Cataloguing in Publication Data
A catalogue record for this book is available from the British Library

ISBN 978 1 292244 81 5

Printed in Slovakia by Neografia

Picture Credits
The authors and publisher would like to thank the following individuals and organisations for permission to reproduce photographs:

Shutterstock.com: Vladi333 1, Vitalii Nesterchuk 20; **Getty Images:** Andy Fritzsche/EyeEm 41; **Alamy Stock Photo:** SJ Images 85, Nano Calvo 120, **123RF:** jarre. 148

Cover images: *Front*: **Getty Images:** Werner Van Steen
Inside front cover: **Shutterstock.com:** Dmitry Lobanov

All other images © Pearson Education Limited 2019
All artwork © Pearson Education Limited 2019

Endorsement Statement
In order to ensure that this resource offers high-quality support for the associated Pearson qualification, it has been through a review process by the awarding body. This process confirms that this resource fully covers the teaching and learning content of the specification or part of a specification at which it is aimed. It also confirms that it demonstrates an appropriate balance between the development of subject skills, knowledge and understanding, in addition to preparation for assessment.

Endorsement does not cover any guidance on assessment activities or processes (e.g. practice questions or advice on how to answer assessment questions) included in the resource, nor does it prescribe any particular approach to the teaching or delivery of a related course.

While the publishers have made every attempt to ensure that advice on the qualification and its assessment is accurate, the official specification and associated assessment guidance materials are the only authoritative source of information and should always be referred to for definitive guidance.

Pearson examiners have not contributed to any sections in this resource relevant to examination papers for which they have responsibility.

Examiners will not use endorsed resources as a source of material for any assessment set by Pearson. Endorsement of a resource does not mean that the resource is required to achieve this Pearson qualification, nor does it mean that it is the only suitable material available to support the qualification, and any resource lists produced by the awarding body shall include this and other appropriate resources.

CONTENTS

ABOUT THIS BOOK

The following three themes have been fully integrated throughout the Pearson Edexcel International Advanced Level in Mathematics series, so they can be applied alongside your learning.

1. Mathematical argument, language and proof

- Rigorous and consistent approach throughout
- Notation boxes explain key mathematical language and symbols

2. Mathematical problem-solving

- Hundreds of problem-solving questions, fully integrated into the main exercises
- Problem-solving boxes provide tips and strategies
- Challenge questions provide extra stretch

The Mathematical Problem-Solving Cycle

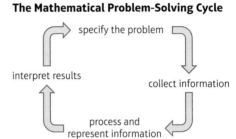

specify the problem

collect information

process and represent information

interpret results

3. Transferable skills

- Transferable skills are embedded throughout this book: in the exercises and in some examples
- These skills are signposted to show students which skills they are using and developing

Finding your way around the book

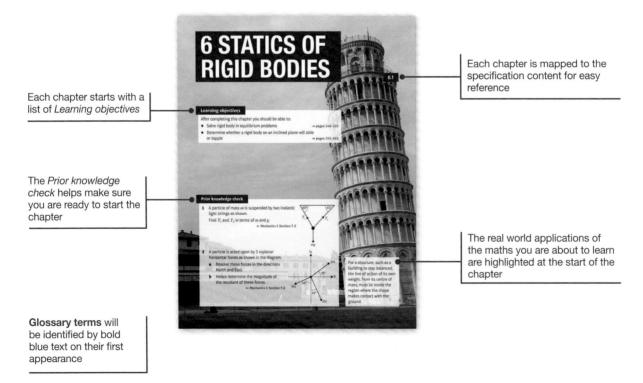

Each chapter is mapped to the specification content for easy reference

Each chapter starts with a list of *Learning objectives*

The *Prior knowledge check* helps make sure you are ready to start the chapter

The real world applications of the maths you are about to learn are highlighted at the start of the chapter

Glossary terms will be identified by bold blue text on their first appearance

Each section begins with explanation and key learning points

Transferable skills are signposted where they naturally occur in the exercises and examples

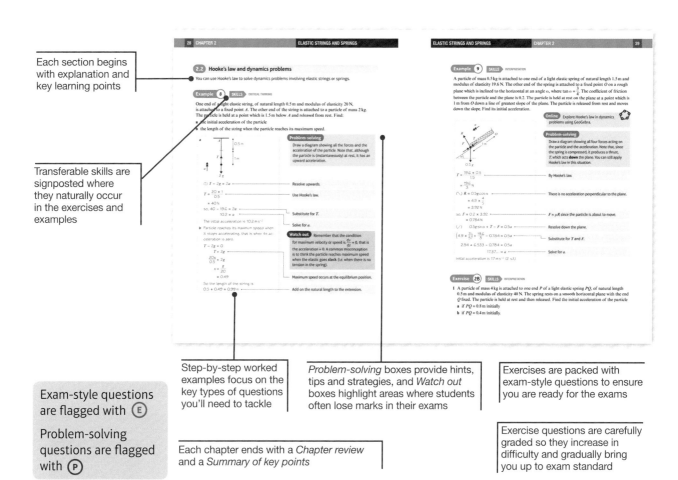

Step-by-step worked examples focus on the key types of questions you'll need to tackle

Problem-solving boxes provide hints, tips and strategies, and *Watch out* boxes highlight areas where students often lose marks in their exams

Exercises are packed with exam-style questions to ensure you are ready for the exams

Exam-style questions are flagged with Ⓔ

Problem-solving questions are flagged with Ⓟ

Each chapter ends with a *Chapter review* and a *Summary of key points*

Exercise questions are carefully graded so they increase in difficulty and gradually bring you up to exam standard

After every few chapters, a *Review exercise* helps you consolidate your learning with lots of exam-style questions

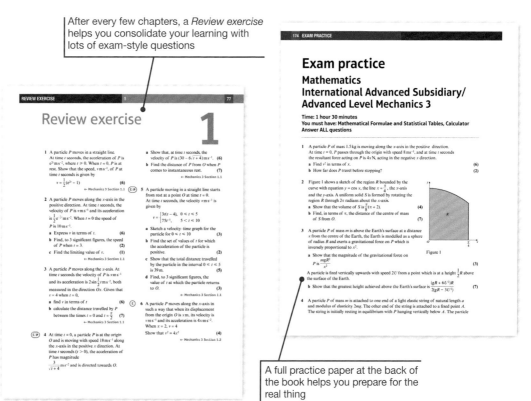

A full practice paper at the back of the book helps you prepare for the real thing

QUALIFICATION AND ASSESSMENT OVERVIEW

Qualification and content overview

Mechanics 3 (M3) is an **optional** unit in the following qualifications:

International Advanced Subsidiary in Further Mathematics

International Advanced Level in Further Mathematics

Assessment overview

The following table gives an overview of the assessment for this unit.

We recommend that you study this information closely to help ensure that you are fully prepared for this course and know exactly what to expect in the assessment.

Unit	Percentage	Mark	Time	Availability
M3: Mechanics 3	$33\frac{1}{3}$ % of IAS	75	1 hour 30 mins	January and June
Paper code WME03/01	$16\frac{2}{3}$ % of IAL			First assessment June 2020

IAS: International Advanced Subsidiary, IAL: International Advanced A Level.

Assessment objectives and weightings

		Minimum weighting in IAS and IAL
AO1	Recall, select and use their knowledge of mathematical facts, concepts and techniques in a variety of contexts.	30%
AO2	Construct rigorous mathematical arguments and proofs through use of precise statements, logical deduction and inference and by the manipulation of mathematical expressions, including the construction of extended arguments for handling substantial problems presented in unstructured form.	30%
AO3	Recall, select and use their knowledge of standard mathematical models to represent situations in the real world; recognise and understand given representations involving standard models; present and interpret results from such models in terms of the original situation, including discussion of the assumptions made and refinement of such models.	10%
AO4	Comprehend translations of common realistic contexts into mathematics; use the results of calculations to make predictions, or comment on the context; and, where appropriate, read critically and comprehend longer mathematical arguments or examples of applications.	5%
AO5	Use contemporary calculator technology and other permitted resources (such as formulae booklets or statistical tables) accurately and efficiently; understand when not to use such technology, and its limitations. Give answers to appropriate accuracy.	5%

Relationship of assessment objectives to units

M3	Assessment objective				
	AO1	AO2	AO3	AO4	AO5
Marks out of 75	20–25	25–30	10–15	5–10	5–10
%	$26\frac{2}{3}$–$33\frac{1}{3}$	$33\frac{1}{3}$–40	$13\frac{1}{3}$–20	$6\frac{2}{3}$–$13\frac{1}{3}$	$6\frac{2}{3}$–$13\frac{1}{3}$

Calculators

Students may use a calculator in assessments for these qualifications. Centres are responsible for making sure that calculators used by their students meet the requirements given in the table below.

Students are expected to have available a calculator with at least the following keys: $+, -, \times, \div, \pi, x^2$, $\sqrt{x}, \frac{1}{x}, x^y, \ln x, e^x, x!$, sine, cosine and tangent and their inverses in degrees and decimals of a degree, and in radians; memory.

Prohibitions

Calculators with any of the following facilities are prohibited in all examinations:

- databanks
- retrieval of text or formulae
- built-in symbolic algebra manipulations
- symbolic differentiation and/or integration
- language translators
- communication with other machines or the internet

Extra online content

Whenever you see an *Online* box, it means that there is extra online content available to support you.

SolutionBank

SolutionBank provides worked solutions for questions in the book. Download the solutions as a PDF or quickly find the solution you need online.

Use of technology

Explore topics in more detail, visualise problems and consolidate your understanding. Use pre-made GeoGebra activities or Casio resources for a graphic calculator.

Online Find the point of intersection graphically using technology.

GeoGebra

GeoGebra-powered interactives

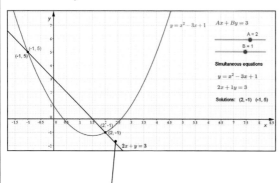

Interact with the maths you are learning using GeoGebra's easy-to-use tools

CASIO®

Graphic calculator interactives

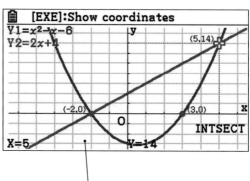

Explore the maths you are learning and gain confidence in using a graphic calculator

---

Calculator tutorials

Our helpful video tutorials will guide you through how to use your calculator in the exams. They cover both Casio's scientific and colour graphic calculators.

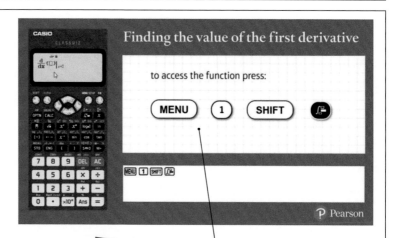

Online Work out each coefficient quickly, using the ^nC_r and power functions on your calculator.

Step-by-step guide with audio instructions on exactly which buttons to press and what should appear on your calculator's screen

1 KINEMATICS

Learning objectives

After completing this chapter you should be able to:

● Use calculus with a particle moving in a straight line and with
   acceleration varying with time   → pages 1-10

● Use calculus with a particle moving in a straight line and with
   acceleration varying with displacement   → pages 10-19

Prior knowledge check

1 Integrate with respect to x

   a $\dfrac{8}{(2-3x)^3}$ **b** $4e^{3x}$ **c** $\sin 5\pi x$

   ← **Pure 3 Section 6.1**

2 $\dfrac{dy}{dx} = \dfrac{y}{(x+2)^2}$

   Find y in terms of x given $y = 3$ when
   $x = 1$ ← **Pure 4 Section 6.6**

3 Given that $\displaystyle\int_1^2 \dfrac{1}{x^2 + x}\,dx = \ln k$, find k,
   where k is a rational constant to be found.

   ← **Pure 4 Section 6.5**

When an object moves under a force
field, the force acting might change
as the object changes position. You
can use differential equations to
solve problems where acceleration
is a function of position.

1.1 Acceleration varying with time

You can use calculus for a **particle** moving in a straight line with **acceleration** that varies with time.

- To find the **velocity** from the **displacement**, you differentiate with respect to time.
  To find the acceleration from the velocity, you differentiate with respect to time.

$$v = \frac{dx}{dt} \text{ and } a = \frac{dv}{dt} = \frac{d^2x}{dt^2}$$

- To obtain the velocity from the acceleration, you integrate with respect to time.
  To obtain the displacement from the velocity, you integrate with respect to time.

$$v = \int a\,dt \text{ and } x = \int v\,dt$$

These relationships are summarised in the following diagram.

Displacement (x)

Differentiate ↓   ↑ Integrate

Velocity (v)

Differentiate ↓   ↑ Integrate

Acceleration (a)

Watch out When you integrate, it is important that you remember to include a **constant** of integration. Many questions include information that enables you to find the value of this constant.

Links If you are given $a = f(t)$ you can use direct integration to find expressions for v or x in terms of t.   ← **Mechanics 2 Section 2.2**

Example **1**   **SKILLS**   PROBLEM-SOLVING

A particle P starts from rest at a point O and moves along a straight line. At time t seconds the acceleration, $a\,\text{m s}^{-2}$, of P is given by

$$a = \frac{6}{(t + 2)^2} \qquad t \geqslant 0$$

a Find the velocity of P at time t seconds.

b Show that the displacement of P from O when $t = 6$ is $(18 - 12\ln 2)\,\text{m}$.

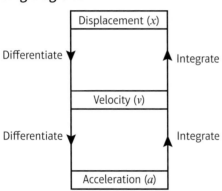

a $a = 6(t + 2)^{-2}$

$v = \int a\,dt = \int 6(t + 2)^{-2}\,dt = \frac{6(t + 2)^{-1}}{-1} + A$

$\quad = A - \frac{6}{t + 2}$

When $t = 0$, $v = 0$

$0 = A - \frac{6}{2} \Rightarrow A = 3$

$v = 3 - \frac{6}{t + 2}$

To integrate, write $\frac{1}{(t + 2)^2}$ as $(t + 2)^{-2}$

Find the velocity by integrating the acceleration with respect to time. It is important to include a constant of integration. The question includes information that enables you to find this constant.

P starts from rest. This means that $v = 0$ when $t = 0$. This initial condition enables you to find the constant of integration.

The velocity of P at time t seconds is

$\left(3 - \dfrac{6}{t+2}\right) m s^{-1}$

b Let the displacement of P from O at time t seconds be s metres.

$s = \int v \, dt = \int \left(3 - \dfrac{6}{t+2}\right) dt$

$= 3t - 6\ln(t+2) + B$

When $t = 0$, $s = 0$

$0 = -6\ln 2 + B \Rightarrow B = 6\ln 2$

$s = 3t - 6\ln(t+2) + 6\ln 2$

When $t = 6$

$s = 18 - 6\ln 8 + 6\ln 2$

$= 18 - 6\ln\left(\dfrac{8}{2}\right) = 18 - 6\ln 4 = 18 - 12\ln 2$

The displacement of P from O when $t = 6$ is $(18 - 12\ln 2)$ m, as required.

Find the displacement by integrating the velocity with respect to time. Use a different letter for the constant of integration.

As P starts at O, $s = 0$ when $t = 0$. This enables you to find the second constant of integration.

Use the laws of logarithms to simplify your answer into the form asked for in the question. This can be done in more than one way. The working shown here uses

$\ln 8 - \ln 2 = \ln\left(\dfrac{8}{2}\right) = \ln 4$

and

$\ln 4 = \ln 2^2 = 2\ln 2$

← Pure 2 Section 5.4

Example **2**   **SKILLS** PROBLEM-SOLVING

A particle P is moving along the x-axis. At time $t = 0$, the particle is at the **origin** O and is moving with speed $2 \, m\,s^{-1}$ in the direction Ox. At time t seconds, where $t \geqslant 0$, the acceleration of P is $4e^{-0.5t} \, m\,s^{-2}$ directed away from O.

a Find the velocity of P at time t seconds.

b Show that the speed of P cannot exceed $10 \, m\,s^{-1}$.

c Sketch a velocity–time graph to **illustrate** the motion of P.

a $a = 4e^{-0.5t}$

Let the velocity of P at time t seconds be $v \, m\,s^{-1}$.

$v = \int a \, dt = \int 4e^{-0.5t} \, dt$

$= -8e^{-0.5t} + C$

When $t = 0$, $v = 2$

$2 = -8 + C \Rightarrow C = 10$

$v = 10 - 8e^{-0.5t}$

The velocity of P at time t seconds is $(10 - 8e^{-0.5t}) \, m\,s^{-1}$.

b For all x, $e^x > 0$ and so for all t, $8e^{-0.5t} > 0$

It follows that $10 - 8e^{-0.5t} < 10$ for all t.

Hence, the speed of P cannot exceed $10 \, m\,s^{-1}$.

Use the rule $\int e^{kt} \, dt = \dfrac{1}{k}e^{kt} + C$

Remember to include a constant of integration.

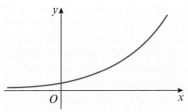

This sketch of $y = e^x$ illustrates that $e^x > 0$, for all real values of x; both positive and negative.

10 minus a positive number must be less than 10.

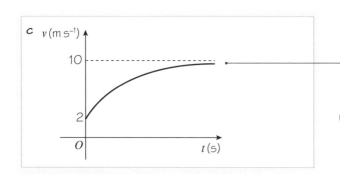

The curve approaches the line $v = 10$ but does not reach it. The line is an **asymptote** to the curve.

Notation

The velocity 10 m s^{-1} is the **terminal** or **limiting velocity** of P.

Example **3**   **SKILLS** PROBLEM-SOLVING

A particle P is moving along the positive x-axis.
At time t seconds, the velocity of P is $v \text{ m s}^{-1}$ in the direction of x increasing, where $v = 4 \sin (2\pi t)$. When $t = 0$, P is at O. Find

a the **magnitude** of the acceleration of P when $t = \dfrac{2}{3}$

b the greatest distance from O attained by P during the motion.

a Let the acceleration of P at time t seconds be $a \text{ m s}^{-2}$.

$$a = \frac{dv}{dt} = 8\pi \cos (2\pi t)$$

When $t = \dfrac{2}{3}$

$$a = 8\pi \cos \left(\frac{4\pi}{3} \right) = 8\pi \times -\frac{1}{2} = -4\pi$$

The magnitude of the acceleration of P when $t = \dfrac{2}{3}$ is $4\pi \text{ m s}^{-2}$.

Find the acceleration by differentiating the velocity with respect to time.

$$\frac{d}{dt} (\sin kt) = k \cos kt$$

Here $k = 2\pi$

When differentiating and integrating trigonometric functions, angles will always be measured in **radians**: $\cos \left(\dfrac{4\pi}{3} \right) = -\dfrac{1}{2}$

b Let the displacement of P at time t seconds be x metres.

$$x = \int v \, dt = -\frac{4}{2\pi} \cos (2\pi t) + C$$

$$= -\frac{2}{\pi} \cos (2\pi t) + C$$

When $t = 0$, $x = 0$

$$0 = -\frac{2}{\pi} + C \Rightarrow C = \frac{2}{\pi}$$

$$x = \frac{2}{\pi} (1 - \cos (2\pi t))$$

The greatest value of x occurs when $\cos (2\pi t) = -1$

The greatest value of x is $\dfrac{2}{\pi} (1 - (-1)) = \dfrac{4}{\pi}$

The greatest distance from O attained by P during the motion is $\dfrac{4}{\pi}$ m.

Find the displacement by integrating the velocity with respect to time. Use the initial condition in the question to find the constant of integration.

When $t = 0$,
$\cos (2\pi t) = \cos 0 = 1$

The cosine of any function varies between $+1$ and -1 and so $1 - \cos (2\pi t)$ varies between 0 and 2. Its greatest value is therefore 2. You do not need to use calculus in this part of the question.

Example 4   SKILLS   PROBLEM-SOLVING

A particle P is moving along the x-axis. **Initially** P is at the origin O.
At time t seconds (where $t \geqslant 0$) the velocity, $v\,\text{m s}^{-1}$, of P is given by $v = t\mathrm{e}^{-\frac{t}{4}}$
Find the distance of P from O when the acceleration of P is zero.

$a = \dfrac{\mathrm{d}v}{\mathrm{d}t} = \mathrm{e}^{-\frac{t}{4}} - \dfrac{1}{4}t\mathrm{e}^{-\frac{t}{4}} = \mathrm{e}^{-\frac{t}{4}}\left(1 - \dfrac{1}{4}t\right)$

When $a = 0$, as $\mathrm{e}^{-\frac{t}{4}} \neq 0$

$\qquad 1 - \dfrac{1}{4}t = 0 \Rightarrow t = 4$

$\qquad x = \displaystyle\int v\,\mathrm{d}t = \int t\mathrm{e}^{-\frac{t}{4}}\,\mathrm{d}t$

$\qquad\quad = -4t\mathrm{e}^{-\frac{t}{4}} + \displaystyle\int 4\mathrm{e}^{-\frac{t}{4}}\,\mathrm{d}t$

$\qquad\quad = -4t\mathrm{e}^{-\frac{t}{4}} - 16\mathrm{e}^{-\frac{t}{4}} + A$

$\qquad\quad = A - \mathrm{e}^{-\frac{t}{4}}(4t + 16)$

When $t = 0$, $x = 0$

$\qquad 0 = A - 16 \Rightarrow A = 16$

Hence

$\qquad x = 16 - \mathrm{e}^{-\frac{t}{4}}(4t + 16)$

When $t = 4$

$\qquad x = 16 - \mathrm{e}^{-1}(4 \times 4 + 16) = 16(1 - 2\mathrm{e}^{-1})$

When the acceleration of P is zero,

$OP = 16(1 - 2\mathrm{e}^{-1})\,\text{m}$

The first step is to find the value of t for which the acceleration is zero. Find the acceleration by differentiating the velocity using the product rule

$\dfrac{\mathrm{d}}{\mathrm{d}t}(uv) = v\dfrac{\mathrm{d}u}{\mathrm{d}t} + u\dfrac{\mathrm{d}v}{\mathrm{d}t}$

with $u = t$ and $v = \mathrm{e}^{-\frac{t}{4}}$

Find the displacement by integrating the velocity with respect to time. You need to use integration by parts

$\displaystyle\int u\dfrac{\mathrm{d}v}{\mathrm{d}t}\,\mathrm{d}t = uv - \int v\dfrac{\mathrm{d}u}{\mathrm{d}t}\,\mathrm{d}t$

with $u = t$ and $\dfrac{\mathrm{d}v}{\mathrm{d}t} = \mathrm{e}^{-\frac{t}{4}}$

Use the information that P is initially at the origin to find the value of the constant of integration A.

Substitute $t = 4$ into your expression for the displacement.

Example 5   SKILLS   PROBLEM-SOLVING

A particle is moving along the x-axis. At time t seconds the velocity of P is $v\,\text{m s}^{-1}$ in the direction of x **increasing**, where

$$v = \begin{cases} 2t & 0 \leqslant t \leqslant 2 \\ 2 + \dfrac{4}{t} & t > 2 \end{cases}$$

When $t = 0$, P is at the origin O.

a Sketch a velocity–time graph to illustrate the motion of P in the **interval** $0 \leqslant t \leqslant 5$

b Find the distance of P from O when $t = 5$

a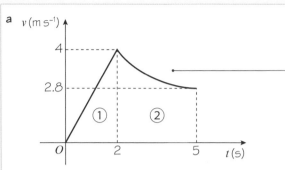

For $v = 2t$, when $t = 0$, $v = 0$ and when $t = 2$, $v = 4$. The graph is the line segment joining $(0, 0)$ to $(2, 4)$.

For $v = 2 + \dfrac{4}{t}$, when $t = 2$, $v = 4$

and when $t = 5$, $v = 2.8$. The graph is part of a reciprocal curve joining $(2, 4)$ to $(5, 2.8)$.

b   The distance moved in the first two seconds is represented by the area labelled ①

Let this area be A_1

$$A_1 = \frac{1}{2} \times 2 \times 4 = 4$$

The distance travelled in the next three seconds is represented by the area labelled ②

Let this area be A_2

$$A_2 = \int_2^5 \left(2 + \frac{4}{t}\right) dt$$

$$= [2t + 4\ln t]_2^5$$

$$= (10 + 4\ln 5) - (4 + 4\ln 2)$$

$$= 6 + 4(\ln 5 - \ln 2) = 6 + 4\ln\frac{5}{2}$$

The distance of P from O when $t = 5$ is

$$4 + 6 + 4\ln\frac{5}{2} = \left(10 + 4\ln\frac{5}{2}\right) m$$

The distance moved by P is represented by the area between the graph and the t-axis.

The area labelled ① can be found using the formula:

area of a triangle $= \dfrac{1}{2} \times$ base $\times$ height.

The area labelled ② in the diagram can be found by definite integration.

Integrate the function $2 + \dfrac{4}{t}$ between the limits $t = 2$ and $t = 5$

Example **6**   **SKILLS**   PROBLEM-SOLVING

A particle P moves on the positive x-axis.

The velocity of P at time t seconds is $(2t^2 - 7t + 3)\,\text{m s}^{-1}$, $t \geqslant 0$

When $t = 0$, P is $10\,\text{m}$ from the origin O.

Find

a   the values of t when P is **instantaneously** at rest

b   the displacement of P from O when $t = 5$

c   the total distance travelled by P in the interval $0 \leqslant t \leqslant 5$

a $v = 2t^2 - 7t + 3$

$2t^2 - 7t + 3 = 0$

$(2t - 1)(t - 3) = 0$ so P is instantaneously at rest when $t = 0.5$ and $t = 3$

Set $v = 0$ to find the times when the particle is instantaneously at rest.

P is instantaneously at rest when $v = 0$

b $s = \int (2t^2 - 7t + 3)\,dt = \dfrac{2}{3}t^3 - \dfrac{7}{2}t^2 + 3t + C$

Integrate the expression for velocity to obtain the displacement.

When $t = 0$, $s = 10$ so $C = 10$

Use the initial conditions given to find the value of C.

$s = \dfrac{2}{3}t^3 - \dfrac{7}{2}t^2 + 3t + 10$

When $t = 5$:

$s = \left(\dfrac{2}{3} \times 5^3\right) - \left(\dfrac{7}{2} \times 5^2\right) + (3 \times 5) + 10 = \dfrac{125}{6} = 20.83\,\text{m}$

This is the expression for the **displacement** of P from the origin at time t.

c Velocity–time graph for the motion of the particle:

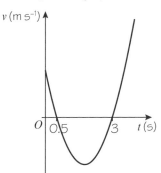

$s = \int (2t^2 - 7t + 3)\,dt$

Between $t = 0$ and $t = \dfrac{1}{2}$

$\left[\dfrac{2}{3}t^3 - \dfrac{7}{2}t^2 + 3t\right]_0^{\frac{1}{2}}$

$= \left(\dfrac{2}{3} \times \left(\dfrac{1}{2}\right)^3 - \dfrac{7}{2} \times \left(\dfrac{1}{2}\right)^2 + 3 \times \dfrac{1}{2}\right) - 0 = \dfrac{17}{24}$

Between $t = \dfrac{1}{2}$ and $t = 3$

$\left[\dfrac{2}{3}t^3 - \dfrac{7}{2}t^2 + 3t\right]_{\frac{1}{2}}^3$

$= \left(\left(\dfrac{2}{3} \times 3^3\right) - \left(\dfrac{7}{2} \times 3^2\right) + (3 \times 3)\right)$

$\quad - \left(\dfrac{2}{3} \times \left(\dfrac{1}{2}\right)^3 - \dfrac{7}{2} \times \left(\dfrac{1}{2}\right)^2 + 3 \times \dfrac{1}{2}\right)$

$= \left(-\dfrac{9}{2} - \dfrac{17}{24}\right) = -\dfrac{125}{24}$

Between $t = 3$ and $t = 5$

$\left[\dfrac{2}{3}t^3 - \dfrac{7}{2}t^2 + 3t\right]_3^5$

$= \left(\left(\dfrac{2}{3} \times 5^3\right) - \left(\dfrac{7}{2} \times 5^2\right) + (3 \times 5)\right)$

$\quad - \left(\left(\dfrac{2}{3} \times 3^3\right) - \left(\dfrac{7}{2} \times 3^2\right) + (3 \times 3)\right)$

$= \dfrac{65}{6} - \left(-\dfrac{9}{2}\right) = \dfrac{46}{3}$

Total distance travelled by P in the interval $0 \leqslant t \leqslant 5$

$\dfrac{17}{24} + \dfrac{125}{24} + \dfrac{46}{3} = 21.25\,\text{m}$

Problem-solving

The particle **changes direction** twice in the interval $0 \leqslant t \leqslant 5$

If you were to find $\int_0^5 v\,dt$ you would be working out the **displacement** of the particle at time $t = 5$ from its position at time $t = 0$

To work out the **distance travelled** you need to find the total area enclosed (i.e. surrounded) by the velocity–time graph and the x-axis. Sketch the graph to show the critical points, and work out three separate integrals.

Distance travelled

$= \int_0^{0.5} v\,dt - \int_{0.5}^3 v\,dt + \int_3^5 v\,dt$

The negative term arises (i.e. starts to exist) because the definite integral will be negative for an area below the x-axis.

The distance travelled between $t = \dfrac{1}{2}$ and $t = 3$ is $\dfrac{125}{24}\,\text{m}$

Exercise (1A)   **SKILLS**   PROBLEM-SOLVING

1 A particle P is moving in a straight line. Initially P is moving through a point O with speed $4\,\text{m s}^{-1}$. At time t seconds after passing through O the acceleration of P is $3e^{-0.25t}\,\text{m s}^{-2}$ in the direction OP. Find the velocity of the particle at time t seconds.

2 A particle P is moving along the x-axis in the direction of x increasing.
At time t seconds, the velocity of P is $t\sin t\,\text{m s}^{-1}$. When $t = 0$, P is at the origin.
Show that when $t = \dfrac{\pi}{2}$, P is 1 metre from O.

3 At time t seconds the velocity, $v\,\text{m s}^{-1}$, of a particle moving in a straight line is given by

$$v = \frac{4}{3 + 2t} \qquad t \geqslant 0$$

When $t = 0$, the particle is at a point A. When $t = 3$, the particle is at the point B. Find the distance between A and B.

4 A particle P is moving along the x-axis in the positive direction. At time t seconds the acceleration of P is $4e^{\frac{1}{2}t}\,\text{m s}^{-2}$ in the positive direction. When $t = 0$, P is at rest. Find the distance P moves in the interval $0 \leqslant t \leqslant 2$. Give your answer to 3 significant figures (s.f.).

5 A particle P is moving along the x-axis. At time t seconds the displacement of P from O is x m and the velocity of P is $4\cos 3t\,\text{m s}^{-1}$, both measured in the direction Ox. When $t = 0$ the particle P is at the origin O. Find

  a the magnitude of the acceleration when $t = \dfrac{\pi}{12}$

  b x in terms of t

  c the smallest positive value of t for which P is at O.

6 A particle P is moving along a straight line. Initially P is at rest. At time t seconds P has velocity $v\,\text{m s}^{-1}$ and acceleration $a\,\text{m s}^{-2}$ where

$$a = \frac{6t}{(2 + t^2)^2} \qquad t \geqslant 0$$

Find v in terms of t.

(P) **7** A particle P is moving along the x-axis. At time t seconds the velocity of P is $v\,\text{m s}^{-1}$ in the direction of x increasing, where

$$v = \begin{cases} 4 & 0 \leqslant t \leqslant 3 \\ 5 - \dfrac{3}{t} & 3 < t \leqslant 6 \end{cases}$$

When $t = 0$, P is at the origin O.

  a Sketch a velocity–time graph to illustrate the motion of P in the interval $0 \leqslant t \leqslant 6$

  b Find the displacement of P from O when $t = 6$

8 A particle P is moving in a straight line with acceleration $\sin\frac{1}{2}t$ m s^{-2} at time t seconds, $t \geqslant 0$. The particle is initially at rest at a point O. Find

    a the speed of P when $t = 2\pi$

    b the displacement of P from O when $t = \dfrac{\pi}{2}$

(E/P) **9** A particle P is moving along the x-axis. At time t seconds P has velocity v m s^{-1} in the direction x increasing and an acceleration of magnitude $4e^{0.2t}$ m s^{-2} in the direction x **decreasing**. When $t = 0$, P is moving through the origin with velocity 20 m s^{-1} in the direction x increasing. Find

    a v in terms of t **(3 marks)**

    b the **maximum** value of x attained by P during its motion.   **(3 marks)**

(P) **10** A car is travelling along a straight road. As it passes a sign S, the driver applies the brakes. The car is **modelled** as a particle. At time t seconds the car is x m from S and its velocity, v m s^{-1}, is modelled by the equation $v = \dfrac{3200}{c + kt}$ where c and k are constants.

    Given that when $t = 0$, the speed of the car is 40 m s^{-1} and its **deceleration** is 0.5 m s^{-2}, find

    a the value of c and the value of k

    b x in terms of t.

(P) **11** A particle P is moving along a straight line. When $t = 0$, P is passing through a point A. At time t seconds after passing through A the velocity, v m s^{-1}, of P is given by

        $v = e^{2t} - 11e^t + 15t$

    Find

    a the values of t for which the acceleration is zero

    b the displacement of P from A when $t = \ln 3$

(P) **12** A particle P moves along a straight line. At time t seconds (where $t > 0$) the velocity of P is $(2t + \ln(t + 2))$ m s^{-1}. Find

    a the value of t for which the acceleration has magnitude 2.2 m s^{-2}

    b the distance moved by P in the interval $1 \leqslant t \leqslant 4$

(E/P) **13** A particle P moves on the positive x-axis.

    The velocity of P at time t seconds is $(3t^2 - 5t + 2)$ m s^{-1}, $t \geqslant 0$

    When $t = 0$, P is at the origin O.

    Find

    a the values of t when P is instantaneously at rest   **(2 marks)**

    b the acceleration of P when $t = 5$ **(3 marks)**

    c the total distance travelled by P in the interval $0 \leqslant t \leqslant 5$ **(5 marks)**

    When $t = 0$, P is at the origin O.

    d Show that P never returns to O,

        explaining your reasoning.   **(3 marks)**

> **Problem-solving**
>
> In part **d**, form an expression for the displacement and show that $d \neq 0$ for any value of t except $t = 0$

(E/P) **14** A particle moving in a straight line starts from rest at the point O at time $t = 0$
At time t seconds, the velocity $v\,\text{m s}^{-1}$ of the particle is given by

$$v = 2t(t - 5) \qquad 0 \leqslant t \leqslant 6$$

$$v = \frac{72}{t} \qquad 6 < t \leqslant 12$$

   a Sketch a velocity–time graph for the particle for $0 \leqslant t \leqslant 12$ **(3 marks)**

   b Find the set of values of t for which the acceleration of the particle is positive.   **(2 marks)**

   c Find the total distance travelled by the particle in the interval $0 \leqslant t \leqslant 12$ **(5 marks)**

Challenge

A truck travels along a straight road. The truck is modelled as a particle.

At time t seconds, $t \geqslant 2$, the acceleration is given by $\dfrac{60}{kt^2}\,\text{m s}^{-2}$ where k is a positive constant.

When $t = 2$ the truck is at rest and when $t = 5$ the speed of the truck is $9\,\text{m s}^{-1}$.

Show that the speed of the truck never reaches $15\,\text{m s}^{-1}$.

1.2 Acceleration varying with displacement

You can use calculus for a particle moving in a straight line with acceleration that varies with displacement.

When the acceleration of a particle is varying with time, the displacement (x), velocity (v) and acceleration (a) are connected by the relationships

$$a = \frac{\mathrm{d}v}{\mathrm{d}t} = \frac{\mathrm{d}^2 x}{\mathrm{d}t^2}$$

Using the chain rule for differentiation

$$a = \frac{\mathrm{d}v}{\mathrm{d}t} = \frac{\mathrm{d}v}{\mathrm{d}x} \times \frac{\mathrm{d}x}{\mathrm{d}t}$$

As $v = \dfrac{\mathrm{d}x}{\mathrm{d}t}$

$$a = \frac{\mathrm{d}v}{\mathrm{d}x} \times v = v\frac{\mathrm{d}v}{\mathrm{d}x} \qquad \textbf{(1)}$$

Also, if you differentiate $\dfrac{1}{2}v^2$ implicitly with respect to x, you obtain

$$\frac{\mathrm{d}}{\mathrm{d}x}\left(\frac{1}{2}v^2\right) = \frac{1}{2} \times 2v \times \frac{\mathrm{d}v}{\mathrm{d}x} = v\frac{\mathrm{d}v}{\mathrm{d}x} \qquad \textbf{(2)}$$

Combining results **(1)** and **(2)**, you obtain

■ $a = v\dfrac{\mathrm{d}v}{\mathrm{d}x} = \dfrac{\mathrm{d}}{\mathrm{d}x}\left(\dfrac{1}{2}v^2\right)$

You can use these two forms for acceleration to solve problems where the acceleration of a particle varies with displacement.

For example, if you have an equation of the form

$$a = f(x)$$

you can write this as

$$\frac{d}{dx}\left(\frac{1}{2}v^2\right) = f(x)$$

Integrating both sides of the equation with respect to x,

$$\frac{1}{2}v^2 = \int f(x)\, dx$$

Example (7) **SKILLS** CRITICAL THINKING

A particle P is moving on the x-axis in the direction of x increasing. When the displacement of P from the origin O is x m and its speed is v m s^{-1}, the acceleration of P is $2x$ m s^{-2}. When P is at O, its speed is 6 m s^{-1}. Find v in terms of x.

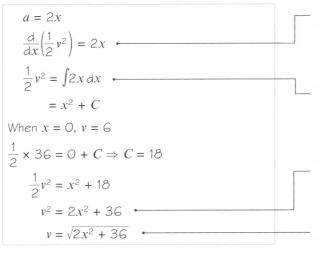

$$a = 2x$$

$$\frac{d}{dx}\left(\frac{1}{2}v^2\right) = 2x$$

$$\frac{1}{2}v^2 = \int 2x\, dx$$

$$= x^2 + C$$

When $x = 0$, $v = 6$

$$\frac{1}{2} \times 36 = 0 + C \Rightarrow C = 18$$

$$\frac{1}{2}v^2 = x^2 + 18$$

$$v^2 = 2x^2 + 36$$

$$v = \sqrt{2x^2 + 36}$$

Use $a = \dfrac{d}{dx}\left(\dfrac{1}{2}v^2\right)$

Integrate both sides of the equation with respect to x. As integration is the inverse process of differentiation,

$$\int \frac{d}{dx}\left(\frac{1}{2}v^2\right) dx = \frac{1}{2}v^2$$

Multiply by 2 throughout the equation and make v the subject of the formula.

As the question tells you that P is moving in the direction of x increasing, you do not need to consider the negative square root.

Example (8) **SKILLS** PROBLEM-SOLVING

A particle P is moving along a straight line. The acceleration of P, when it has displacement x m from a fixed point O on the line and velocity v m s^{-1}, is of magnitude $4x$ m s^{-2} and is directed towards O. At $x = 0$, $v = 20$. Find the values of x for which P is instantaneously at rest.

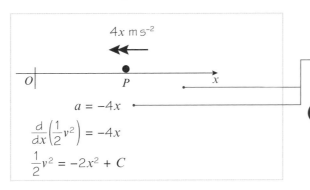

$$4x \text{ m s}^{-2}$$

$$a = -4x$$

$$\frac{d}{dx}\left(\frac{1}{2}v^2\right) = -4x$$

$$\frac{1}{2}v^2 = -2x^2 + C$$

As its acceleration is towards O when P has positive displacement, the acceleration is in the direction of x decreasing, so the acceleration is negative.

Watch out When integrating, you must remember to include a constant of integration. In this question the information that at $x = 0$, $v = 20$ enables you to evaluate the constant.

At $x = 0$, $v = 20$

$\frac{1}{2} \times 20^2 = 0 + C \Rightarrow C = 200$

$\frac{1}{2}v^2 = -2x^2 + 200$

$v^2 = 400 - 4x^2$ •———— The particle is instantaneously at rest when $v = 0$
Substitute $v = 0$ into this expression and solve the resulting equation for x.

When $v = 0$

$0 = 400 - 4x^2 \Rightarrow x^2 = 100$

$x = \pm 10$ •———— There are two points at which $v = 0$
The particle reverses (i.e. changes to become the opposite) direction at these points and will oscillate between them.

The values of x for which P is instantaneously at rest are 10 and -10

Example (9)   **SKILLS**   PROBLEM-SOLVING

A particle P is moving along the positive x-axis in the direction of x increasing. When $OP = x$ m, the velocity of P is v m s^{-1} and the acceleration of P is $\left(\dfrac{54}{x^3} - \dfrac{18}{x^5}\right)$ m s^{-2} where $x \geqslant 1$

Given that $v = 6$ at $x = 1$, find v in terms of x.

$a = \dfrac{54}{x^3} - \dfrac{18}{x^5}$

$\dfrac{d}{dx}\left(\dfrac{1}{2}v^2\right) = \dfrac{54}{x^3} - \dfrac{18}{x^5}$ •———— Integrate both sides of this equation with respect to x using $\int \dfrac{d}{dx}\left(\dfrac{1}{2}v^2\right) dx = \dfrac{1}{2}v^2$ and $\int x^n\, dx = \dfrac{x^{n+1}}{n+1}$
Remember to include a constant of integration.

$= 54x^{-3} - 18x^{-5}$

$\dfrac{1}{2}v^2 = \dfrac{54x^{-2}}{-2} - \dfrac{18x^{-4}}{-4} + C$

$= C - \dfrac{27}{x^2} + \dfrac{9}{2x^4}$

At $x = 1$, $v = 6$

$18 = C - 27 + \dfrac{9}{2} \Rightarrow C = \dfrac{81}{2}$ •———— Multiply throughout by 2 and then factorise the right hand side of the equation.

$\dfrac{1}{2}v^2 = \dfrac{81}{2} - \dfrac{27}{x^2} + \dfrac{9}{2x^4}$ •———— Take the square root of both sides of this equation. As P is moving in the positive direction (the direction of x increasing), you can reject the other square root $v = -\left(9 - \dfrac{3}{x^2}\right)$
This expression is negative for $x \geqslant 1$

$v^2 = 81 - \dfrac{54}{x^2} + \dfrac{9}{x^4} = \left(9 - \dfrac{3}{x^2}\right)^2$

$v = 9 - \dfrac{3}{x^2}$ •————

Example (10)   **SKILLS**   INTERPRETATION

A particle P is moving along the x-axis. Initially P is at the origin O and is moving with velocity 1 m s^{-1} in the direction of x increasing. At time t seconds, P is x m from O, has velocity v m s^{-1} and acceleration of magnitude $\dfrac{1}{2}e^{-x}$ m s^{-2} directed towards O. Find

a v in terms of x

b x in terms of t.

a $\quad a = -\dfrac{1}{2}e^{-x}$

As the acceleration is directed toward O, it is in the direction of x decreasing and is negative.

$$\frac{d}{dx}\left(\frac{1}{2}v^2\right) = -\frac{1}{2}e^{-x}$$

$$\frac{1}{2}v^2 = \frac{1}{2}e^{-x} + A$$

In this example there are two different constants of integration. The initial conditions given in the question enable you to evaluate both constants.

At $x = 0$, $v = 1$

$$\frac{1}{2} = \frac{1}{2} + A \Rightarrow A = 0$$

$$\frac{1}{2}v^2 = \frac{1}{2}e^{-x}$$

The question requires you to make v the subject of the formula.

$$v^2 = e^{-x}$$

$$v = e^{-\frac{x}{2}}$$

b $\quad \dfrac{dx}{dt} = e^{-\frac{x}{2}}$

This is a differential equation of the form $\dfrac{dx}{dt} = f(x)$

You can solve it by separating the **variables**.

← **Pure 4 Section 6.6**

$$e^{\frac{x}{2}}\frac{dx}{dt} = 1$$

$$\int e^{\frac{x}{2}}\,dx = \int 1\,dt$$

$$2e^{\frac{x}{2}} = t + B$$

When $t = 0$, $x = 0$

$$2 = 0 + B \Rightarrow B = 2$$

$$2e^{\frac{x}{2}} = t + 2$$

$$e^{\frac{x}{2}} = \frac{t}{2} + 1$$

To make x the subject of this formula, take logarithms on both sides of the equation and use the property that $\ln e^{f(x)} = f(x)$

$$\frac{x}{2} = \ln\left(\frac{t}{2} + 1\right)$$

$$x = 2\ln\left(\frac{t}{2} + 1\right)$$

Example 11   **SKILLS**   ANALYSIS

A particle P is moving along the positive x-axis. At $OP = x$ m, the velocity of P is v m s^{-1} and the acceleration of P is $\dfrac{k}{(2x + 3)^2}$ m s^{-2}, where k is a constant, directed away from O. At $x = 1$, $v = 10$ and at $x = 6$, $v = \sqrt{120}$

a Find the value of k.

b Show that the speed of P cannot exceed $\sqrt{130}$ m s^{-1}.

a $\quad a = \dfrac{k}{(2x + 3)^2}$

Integrate both sides of this equation with respect to x. For constants a and b and $n \neq -1$

$$\frac{d}{dx}\left(\frac{1}{2}v^2\right) = \frac{k}{(2x + 3)^2}$$

$$\int (ax + b)^n\,dx = \frac{1}{(n + 1)a}(ax + b)^{n+1} + C$$

$$\frac{1}{2}v^2 = A - \frac{k}{2(2x + 3)}$$

$$v^2 = B - \frac{k}{2x + 3}$$

Multiply this equation throughout by 2. Twice one **arbitrary** constant is another arbitrary constant.

At $x = 1$, $v = 10$

$$100 = B - \frac{k}{5} \quad \textbf{(1)}$$

At $x = 6$, $v = \sqrt{120}$

$$120 = B - \frac{k}{15} \quad \textbf{(2)}$$

> The conditions given in the question give you a pair of simultaneous equations in B and k. You find k to solve part **a**. You will also need to find B to solve part **b**.

(2) − **(1)**

$$20 = -\frac{k}{15} - \left(-\frac{k}{5}\right) = \frac{2k}{15}$$

$$k = 20 \times \frac{15}{2} = 150$$

b Substituting $k = 150$ into **(1)**

$$100 = B - \frac{150}{5} \Rightarrow B = 130$$

$$v^2 = 130 - \frac{150}{2x + 3}$$

As x is moving along the positive x-axis $x > 0$, and so both $2x + 3$ and $\frac{150}{2x + 3}$ are positive.

Hence

$$v^2 = 130 - \frac{150}{2x + 3} < 130$$

The speed of P cannot exceed $\sqrt{130}\,\text{m s}^{-1}$.

> As x increases, the velocity of P approaches $\sqrt{130}\,\text{m s}^{-1}$ in the direction Ox asymptotically (i.e. always approaching the curve without ever touching it).

Exercise (1B) SKILLS PROBLEM-SOLVING

1 A particle P moves along the x-axis. At time $t = 0$, P passes through the origin O with velocity $5\,\text{m s}^{-1}$ in the direction of x increasing. At time t seconds, the velocity of P is $v\,\text{m s}^{-1}$ and $OP = x\,\text{m}$. The acceleration of P is $\left(2 + \frac{1}{2}x\right)\text{m s}^{-2}$, measured in the positive x direction. Find v^2 in terms of x.

2 A particle P moves along a straight line. When its displacement from a fixed point O on the line is $x\,\text{m}$ and its velocity is $v\,\text{m s}^{-1}$, the deceleration of P is $4x\,\text{m s}^{-2}$. At $x = 2$, $v = 8$. Find v in terms of x.

3 A particle P is moving along the x-axis in the direction of x increasing. At $OP = x\,\text{m}\ (x > 0)$, the velocity of P is $v\,\text{m s}^{-1}$ and its acceleration is of magnitude $\frac{4}{x^2}\,\text{m s}^{-2}$ in the direction of x increasing. Given that at $x = 2$, $v = 6$, find the value of x for which P is instantaneously at rest.

(P) 4 A particle P moves along a straight line. When its displacement from a fixed point O on the line is $x\,\text{m}$ and its velocity is $v\,\text{m s}^{-1}$ in the direction of x increasing, the acceleration of P is of magnitude $25x\,\text{m s}^{-2}$ and is directed towards O. At $x = 0$, $v = 40$. In its motion P is instantaneously at rest at two points, A and B. Find the distance between A and B.

(P) **5** A particle P is moving along the x-axis. At $OP = x$ m, the velocity of P is v m s^{-1} in the direction of x increasing and its acceleration is of magnitude kx^2 m s^{-2}, where k is a positive constant in the direction of x decreasing. At $x = 0$, $v = 16$. The particle is instantaneously at rest at $x = 20$ Find

   a the value of k

   b the velocity of P when $x = 10$

6 A particle P is moving along the x-axis in the direction of x increasing. At $OP = x$ m, the velocity of P is v m s^{-1} in the direction of x increasing and its acceleration is of magnitude $8x^3$ m s^{-2} in the direction PO. At $x = 2$, $v = 32$. Find the value of x for which $v = 8$

7 A particle P is moving along the x-axis. When the displacement of P from the origin O is x m, the velocity of P is v m s^{-1} in the direction of x increasing and its acceleration is $6 \sin \dfrac{x}{3}$ m s^{-2}. At $x = 0$, $v = 4$. Find

   a v^2 in terms of x

   b the greatest possible speed of P.

(E) **8** A particle P is moving along the x-axis. At $x = 0$, the velocity of P is 2 m s^{-1} in the direction of x increasing. At $OP = x$ m, the velocity of P is v m s^{-1} and its acceleration is $(2 + 3e^{-x})$ m s^{-2}. Find the velocity of P at $x = 3$. Give your answer to 3 significant figures.   **(6 marks)**

(E) **9** A particle P moves away from the origin O along the positive x-axis. The acceleration of P is of magnitude $\dfrac{4}{2x + 1}$ m s^{-2}, where $OP = x$ m, directed towards O. Given that the speed of P at O is 4 m s^{-1}, find

   a the speed of P at $x = 10$ **(4 marks)**

   b the value of x at which P is instantaneously at rest.   **(6 marks)**

   Give your answers to 3 significant figures.

(E) **10** A particle P is moving along the positive x-axis. At $OP = x$ m, the velocity of P is v m s^{-1} and its acceleration is $\left(x - \dfrac{4}{x^3} \right)$ m s^{-2}. The particle starts from the position where $x = 1$ with velocity 3 m s^{-1} in the direction of x increasing. Find

   a v in terms of x **(4 marks)**

   b the least speed of P during its motion.   **(6 marks)**

(E/P) **11** A particle P is moving along the x-axis. Initially P is at the origin O moving with velocity 15 m s^{-1} in the direction of x increasing. When the displacement of P from O is x m, its acceleration is of magnitude $\left(10 + \tfrac{1}{4} x \right)$ m s^{-2} directed towards O. Find the distance P moves before first coming to instantaneous rest.   **(7 marks)**

(E/P) **12** A particle P is moving along the x-axis. At time t seconds, P is x m from O, has velocity v m s^{-1} and acceleration of magnitude $6x^{\frac{1}{3}}$ m s^{-2} in the direction of x increasing. When $t = 0$, $x = 8$ and $v = 12$. Find

   a v in terms of x **(4 marks)**

   b x in terms of t.   **(4 marks)**

Challenge

A particle P moves along the x-axis. At time $t = 0$, P passes through the origin moving in the positive x direction. At time t seconds, the velocity of P is v m s^{-1} and $OP = x$ metres. The acceleration of P is $\frac{1}{10}(25 - x)$

Given that the maximum speed of P is 12 m s^{-1}, find an expression for v^2 in terms of x.

Chapter review (1)   **SKILLS** ▸ PROBLEM-SOLVING

(E) **1** A particle P is moving along the x-axis. At time t seconds, the displacement of P from the origin O is x m and the velocity of P is $4e^{0.5t}$ m s^{-1} in the direction Ox. When $t = 0$, P is at O. Find

    a x in terms of t **(6 marks)**

    b the acceleration of P when $t = \ln 9$ **(3 marks)**

(E) **2** A particle P moves along the x-axis in the direction of x increasing. At time t seconds, the velocity of P is v m s^{-1} and its acceleration is $20te^{-t^2}$ m s^{-2}. When $t = 0$ the speed of P is 8 m s^{-1}. Find

    a v in terms of t **(3 marks)**

    b the limiting velocity of P.   **(2 marks)**

(E) **3** A particle P moves along a straight line. Initially P is at rest at a point O on the line. At time t seconds, where $t \geqslant 0$, the acceleration of P is $\dfrac{18}{(2t + 3)^3}$ m s^{-2} directed away from O.

    Find the value of t for which the speed of P is 0.48 m s^{-1}.   **(4 marks)**

(E) **4** A car moves along a **horizontal** straight road. At time t seconds the acceleration of the car is $\dfrac{100}{(2t + 5)^2}$ m s^{-2} and the velocity is v m s^{-1} in the direction of motion of the car. When $t = 0$, the car is at rest. Find

    a an expression for v in terms of t **(3 marks)**

    b the distance moved by the car in the first 10 seconds of its motion.   **(3 marks)**

(E) **5** A particle P is moving in a straight line with acceleration $\cos^2 t$ m s^{-2} at time t seconds. The particle is initially at rest at a point O.

    a Find the speed of P when $t = \pi$ **(4 marks)**

    b Show that the distance of P from O when $t = \dfrac{\pi}{4}$ is $\frac{1}{64}(\pi^2 + 8)$ m   **(4 marks)**

(E/P) **6** A particle P is moving along the x-axis. At time t seconds, the velocity of P is $v\,\mathrm{m\,s^{-1}}$ in the direction of x increasing, where

$$v = \begin{cases} \frac{1}{2}t^2 & 0 \leqslant t \leqslant 4 \\ 8e^{4-t} & t > 4 \end{cases}$$

When $t = 0$, P is at the origin O. Find

  a the acceleration of P when $t = 2.5$ **(2 marks)**

  b the acceleration of P when $t = 5$ **(2 marks)**

  c the distance of P from O when $t = 6$ **(3 marks)**

(E) **7** A particle P is moving along the x-axis. At time t seconds, P has velocity $v\,\mathrm{m\,s^{-1}}$ in the direction of x increasing and an acceleration of magnitude $\dfrac{2t + 3}{t + 1}\,\mathrm{m\,s^{-2}}$ in the direction of x increasing. When $t = 0$, P is at rest at the origin O. Find

  a v in terms of t **(5 marks)**

  b the distance of P from O when $t = 2$ **(3 marks)**

(E/P) **8** A particle moving in a straight line starts from rest at the point O at time $t = 0$. At time t seconds, the velocity $v\,\mathrm{m\,s^{-1}}$ of the particle is given by

$$v = \begin{cases} 3t^2 - 14t + 8 & 0 \leqslant t \leqslant 5 \\ 18 - \dfrac{t^2}{5} & 5 < t \leqslant T \end{cases}$$

where T is the first time the particle comes to momentary (i.e. instantaneous) rest when travelling with velocity

$18 - \dfrac{t^2}{5}\,\mathrm{m\,s^{-1}}$

  a Find the value of T. **(2 marks)**

  b Sketch a velocity–time graph for the particle for $0 \leqslant t \leqslant T$ **(3 marks)**

  c Find the set of values of t for which the acceleration of the particle is positive. **(2 marks)**

  d Find the total distance travelled by the particle in the interval $0 \leqslant t \leqslant T$ **(5 marks)**

(E/P) **9** A particle P moves on the x-axis. At time t seconds the velocity of P is $v\,\mathrm{m\,s^{-1}}$ in the direction of x increasing, where $v = (t - 4)(3t - 8)$, $t \geqslant 0$

When $t = 0$, P is at the origin O.

  a Find the acceleration of P at time t seconds. **(2 marks)**

  b Find the total distance travelled by P in the first 3 seconds of its motion. **(3 marks)**

  c Show that P never returns to O, explaining your reasoning. **(3 marks)**

(E/P) **10** A particle P moves on the positive x-axis with an acceleration at time t seconds of $(3t - 4)\,\text{m s}^{-2}$. The particle starts from O with a velocity of $2\,\text{m s}^{-1}$.

Find

a the values of t when P is instantaneously at rest   **(4 marks)**

b the total distance travelled by P in the interval $0 \leqslant t \leqslant 4$ **(4 marks)**

(E/P) **11** A particle P moves along a straight line. When the displacement of P from a fixed point on the line is x m, its velocity is $v\,\text{m s}^{-1}$ and its acceleration is of magnitude $\dfrac{6}{x^2}\,\text{m s}^{-2}$ in the direction of x increasing. At $x = 3$, $v = 4$

Find v in terms of x.   **(4 marks)**

(E) **12** A particle is moving along the x-axis. At time $t = 0$, P is passing through the origin O with velocity $8\,\text{m s}^{-1}$ in the direction of x increasing. When P is x m from O, its acceleration is $\left(3 + \dfrac{1}{4}x\right)\text{m s}^{-2}$ in the direction of x decreasing.

Find the positive value of x for which P is instantaneously at rest.   **(5 marks)**

(E) **13** A particle P is moving on the x-axis. When P is a distance x metres from the origin O, its acceleration is of magnitude $\dfrac{15}{4x^2}\,\text{m s}^{-2}$ in the direction OP. Initially P is at the point where $x = 5$ and is moving toward O with speed $6\,\text{m s}^{-1}$.

Find the value of x where P first comes to rest.   **(6 marks)**

(E) **14** A particle P is moving along the x-axis. At time t seconds, the velocity of P is $v\,\text{m s}^{-1}$ and the acceleration of P is $(3 - x)\,\text{m s}^{-2}$ in the direction of x increasing. Initially P is at the origin O and is moving with speed $4\,\text{m s}^{-1}$ in the direction of x increasing. Find

a v^2 in terms of x **(3 marks)**

b the maximum value of v.   **(3 marks)**

(E) **15** A particle P is moving along the x-axis. At time $t = 0$, P passes through the origin O. After t seconds the speed of P is $v\,\text{m s}^{-1}$, $OP = x$ metres and the acceleration of P is $\dfrac{x^2(5 - x)}{2}\,\text{m s}^{-2}$ in the direction of x increasing. At $x = 10$, P is instantaneously at rest. Find

a an expression for v^2 in terms of x **(4 marks)**

b the speed of P when $t = 0$ **(2 marks)**

(E) **16** A particle P moves away from the origin along the positive x-axis. At time t seconds, the acceleration of P is $\dfrac{20}{5x + 2}\,\text{m s}^{-2}$, where $OP = x$ m, directed away from O. Given that the speed of P is $3\,\text{m s}^{-1}$ at $x = 0$, find, giving your answers to 3 significant figures,

a the speed of P at $x = 12$ **(4 marks)**

b the value of x when the speed of P is $5\,\text{m s}^{-1}$.   **(3 marks)**

(E/P) **17** A particle P is moving along the x-axis. When $t = 0$, P is passing through O with velocity 3 m s^{-1} in the direction of x increasing. When $0 \le x \le 4$ the acceleration is of magnitude $\left(4 + \dfrac{1}{2}x\right) \text{ m s}^{-2}$ in the direction of x increasing. At $x = 4$, the acceleration of P changes. For $x > 4$, the magnitude of the acceleration remains $\left(4 + \dfrac{1}{2}x\right) \text{ m s}^{-2}$ but it is now in the direction of x decreasing.

    a Find the speed of P at $x = 4$ **(4 marks)**

    b Find the positive value of x for which P is instantaneously at rest. Give your answer to 2 significant figures.   **(3 marks)**

(E) **18** A particle P is moving along the x-axis. At time t seconds P is x m from O, has velocity $v \text{ m s}^{-1}$ and acceleration of magnitude $(4x + 6) \text{ m s}^{-2}$ in the direction of x increasing. When $t = 0$, P is passing through O with velocity 3 m s^{-1} in the direction of x increasing. Find

    a v in terms of x **(3 marks)**

    b x in terms of t.   **(4 marks)**

Challenge

A rocket is launched straight upwards from the Earth's surface with an initial velocity of $32\,500 \text{ km h}^{-1}$. The flight of the rocket can be modelled as a particle with an acceleration of $-\dfrac{c}{x^2} \text{ km s}^{-2}$ where $c = 4 \times 10^5$ and x km is the distance from the centre of the Earth. The radius of the Earth is 6370 km. Work out the maximum height above the surface of the Earth that the rocket will reach.

Summary of key points

1 To find the velocity from the displacement, you differentiate with respect to time. To find the acceleration from the velocity, you differentiate with respect to time.

$$v = \frac{\mathrm{d}x}{\mathrm{d}t} \text{ and } a = \frac{\mathrm{d}v}{\mathrm{d}t} = \frac{\mathrm{d}^2 x}{\mathrm{d}t^2}$$

2 To obtain the velocity from the acceleration, you integrate with respect to time. To obtain the displacement from the velocity, you integrate with respect to time.

$$v = \int a \, \mathrm{d}t \text{ and } x = \int v \, \mathrm{d}t$$

3 When the acceleration is a function of the displacement you can use

$$a = v\frac{\mathrm{d}v}{\mathrm{d}x} = \frac{\mathrm{d}}{\mathrm{d}x}\left(\frac{1}{2}v^2\right)$$

2 ELASTIC STRINGS AND SPRINGS

Learning objectives

After completing this chapter you should be able to:

● Use Hooke's law to solve equilibrium problems involving elastic
  strings or springs → pages 20–28

● Use Hooke's law to solve dynamics problems involving elastic
  strings or springs → pages 28–30

● Find the energy stored in an elastic string or spring → pages 31–33

● Solve problems involving elastic energy using the principle of
  conservation of mechanical energy and the work–energy principle
  → pages 33–40

Prior knowledge check

1 Three forces act on a particle.
  Given that the particle is in
  equilibrium, calculate the
  exact values of F and $\tan \theta$.

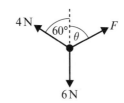

← Mechanics 1 Section 7.1

2 A particle of mass 4 kg is pulled along a rough horizontal table by a
  horizontal force of magnitude 12 N. Given that the mass moves with
  constant velocity, work out the coefficient of friction between the
  particle and the table.

← Mechanics 1 Section 5.3

3 A smooth plane is inclined at 30° to the horizontal. A particle of mass
  0.4 kg slides down a line of greatest slope of the plane. The particle
  starts from rest at point P and passes point Q with a speed 5 m s⁻¹.
  Use the principle of conservation of mechanical energy to find the
  distance PQ. ← Mechanics 2 Section 4.2

Bungee jumping is an activity
that involves jumping from
a high point whilst attached
to a long elastic cord. When
the person jumps, their
gravitational potential energy
is converted (i.e. changed) into
kinetic energy. As the bungee
cord extends, this kinetic
energy (K.E.) is converted into
elastic potential energy.

2.1   Hooke's law and equilibrium problems

You can use Hooke's law to solve **equilibrium** problems involving **elastic** strings or springs.

The **tension** (T) produced when an elastic string or spring is **stretched** is **proportional** to the **extension** (x).

- $T \propto x$
- $T = kx$, where k is a constant

The constant k depends on the unstretched length of the string or spring, l, and the **modulus of elasticity** of the string or spring, λ.

- $T = \dfrac{\lambda x}{l}$

This relationship is called **Hooke's law**.

T is a force measured in newtons, and x and l are both lengths, so the units of λ are also newtons. The value of λ depends on the material from which the elastic string or spring is made, and is a measure of the 'stretchiness' (i.e. how far something can **stretch**) of the string or spring. In this chapter you may assume that Hooke's law applies for the values given in a question. In reality, Hooke's law only applies for values of x up to a maximum value, known as the elastic limit of the spring or string.

> **Watch out**   An elastic spring can also be **compressed**. Instead of a tension this will produce a **thrust** (or **compression**) **force**. Hooke's law still works for compressed elastic springs.

> **Notation**   In this chapter, all elastic strings and springs are modelled as being **light**. This means they have **negligible mass** and do not stretch under their own weight

Example 1   **SKILLS** ▷ PROBLEM-SOLVING

An elastic string of **natural length** 2 m and modulus of elasticity 29.4 N has one end fixed. A particle of mass 4 kg is attached to the other end and hangs at rest. Find the extension of the string.

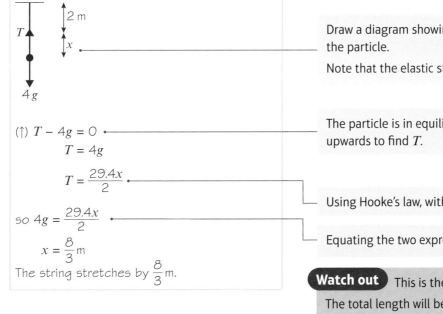

Draw a diagram showing all the forces acting on the particle.

Note that the elastic string is in tension.

($\uparrow$) $T - 4g = 0$

$\qquad T = 4g$

The particle is in equilibrium. **Resolve** vertically upwards to find T.   **← Mechanics 1 Section 5.1**

$T = \dfrac{29.4x}{2}$

Using Hooke's law, with $\lambda = 29.4$ and $l = 2$

so $4g = \dfrac{29.4x}{2}$

Equating the two expressions for T.

$x = \dfrac{8}{3}$ m

The string stretches by $\dfrac{8}{3}$ m.

> **Watch out**   This is the extension in the spring. The total length will be $4\dfrac{2}{3}$ m.

Example **2**   **SKILLS** PROBLEM-SOLVING

An elastic spring of natural length 1.5 m has one end attached to a fixed point. A horizontal force of magnitude 6 N is applied to the other end and compresses the spring to a length of 1 m. Find the modulus of elasticity of the spring.

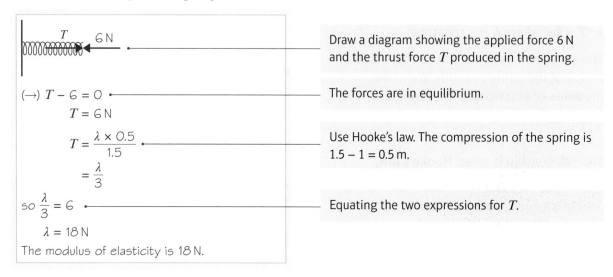

Draw a diagram showing the applied force 6 N and the thrust force T produced in the spring.

$(\rightarrow)\ T - 6 = 0$

$T = 6\,\text{N}$

The forces are in equilibrium.

$T = \dfrac{\lambda \times 0.5}{1.5}$

$= \dfrac{\lambda}{3}$

Use Hooke's law. The compression of the spring is $1.5 - 1 = 0.5\,\text{m}$.

so $\dfrac{\lambda}{3} = 6$

Equating the two expressions for T.

$\lambda = 18\,\text{N}$

The modulus of elasticity is 18 N.

Example **3**   **SKILLS** INTERPRETATION

The elastic springs PQ and QR are joined together at Q to form one long spring. The spring PQ has natural length 1.6 m and modulus of elasticity 20 N. The spring QR has natural length 1.4 m and modulus of elasticity 28 N. The ends, P and R, of the long spring are attached to two fixed points which are 4 m apart, as shown in the diagram.

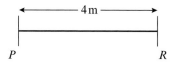

Find the tension in the combined spring.

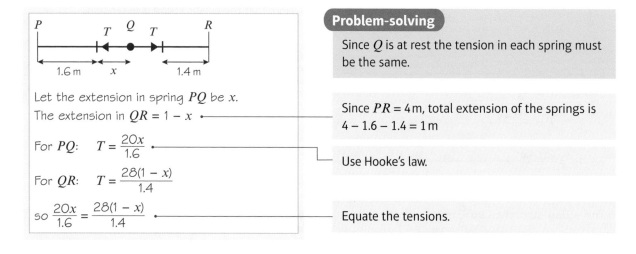

Problem-solving

Since Q is at rest the tension in each spring must be the same.

Let the extension in spring PQ be x.
The extension in $QR = 1 - x$

Since $PR = 4\,\text{m}$, total extension of the springs is $4 - 1.6 - 1.4 = 1\,\text{m}$

For PQ: $T = \dfrac{20x}{1.6}$

Use Hooke's law.

For QR: $T = \dfrac{28(1 - x)}{1.4}$

so $\dfrac{20x}{1.6} = \dfrac{28(1 - x)}{1.4}$

Equate the tensions.

$$\frac{20x}{1.6} = 20(1 - x)$$

$$12.5x = 20 - 20x$$

$$32.5x = 20$$

$$x = \frac{8}{13}$$ ●————————— Solve for x.

so $$T = \frac{20}{1.6} \times \frac{8}{13}$$ ●————————— Substitute for x into tension equation for PQ.

$$= \frac{100}{13} N$$

The tension in the combined spring is

7.69 N (3 s.f.)

Example (4)　　**SKILLS**　ANALYSIS

An elastic string of natural length $2l$ and modulus of elasticity $4mg$ is stretched between two points A and B. The points A and B are on the same horizontal level and $AB = 2l$. A particle P is attached to the midpoint of the string and hangs in equilibrium with both parts of the string making an angle of 30° with the line AB. Find, in terms of m, the mass of the particle.

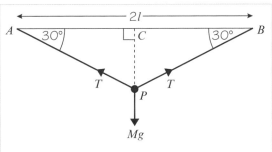

Online　Explore Hooke's law in equilibrium problems involving two elastic springs using GeoGebra.

Problem-solving

Draw a large clear diagram showing the forces acting on the particle. It is useful to label the midpoint of A and B as well.

Let the mass of the particle be M.

(↑) $2T \cos 60° = Mg$ ●————————— The particle is in equilibrium.

$$T = Mg$$

$$AP = \frac{l}{\cos 30°} = \frac{2l}{\sqrt{3}}$$ ●————————— Use $\triangle APC$.

so the stretched length of the string is

$$\frac{4l}{\sqrt{3}}$$ ●————————— Since $AP = PB$

∴ Extension of string is $\left(\dfrac{4l}{\sqrt{3}} - 2l\right)$

∴ $$T = \frac{4mg}{2l}\left(\frac{4l}{\sqrt{3}} - 2l\right)$$ ●————————— Use Hooke's law.

$$= 2mg\left(\frac{4}{\sqrt{3}} - 2\right)$$ ●————————— Cancel the ls.

$$= 0.62mg \ (2 \text{ s.f.})$$

Hence, $0.62mg = Mg$ ●————————— Use $T = Mg$.

The mass of the particle is $0.62m$ (2 s.f.).

Example **5**   **SKILLS**   PROBLEM-SOLVING

An elastic string has natural length 2 m and modulus of elasticity 98 N. One end of the string is attached to a fixed point O and the other end is attached to a particle P of mass 4 kg. The particle is held in equilibrium by a horizontal force of magnitude 28 N, with OP making an angle θ with the **vertical**, as shown. Find

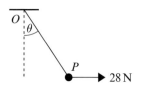

a the value of θ

b the length OP.

Online Explore Hooke's law in equilibrium problems involving one elastic spring using GeoGebra.

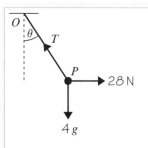

a ($\leftarrow$) $T \sin \theta = 28$

  ($\uparrow$) $T \cos \theta = 4g$

    $\tan \theta = \dfrac{28}{4g} = \dfrac{5}{7}$

so, $\theta = 35.5°$

    $= 36°$ (2 s.f.)

b $T = \dfrac{28}{\sin \theta}$

so $\dfrac{28}{\sin \theta} = \dfrac{98x}{2}$

so $x = \dfrac{4}{7 \sin \theta}$

    $= 0.983...$

$OP = 2 + 0.983... = 2.983...$

$\therefore$ Length of OP is 2.98 m (3 s.f.)

Problem-solving

Since the particle is in equilibrium, you can resolve horizontally and vertically to find θ. You could also use these two equations to find an exact value for T, but it is easier to use your calculator and an unrounded value for θ to find x.

Divide the equations to eliminate T.

Give answer to 2 s.f. as value for g is correct to 2 s.f.

Use Hooke's law, with x as the extension of the string.

Example **6**   **SKILLS**   INTERPRETATION

Two identical elastic springs PQ and QR each have natural length l and modulus of elasticity $2mg$. The springs are joined together at Q. Their other ends, P and R, are attached to fixed points, with P being $4l$ vertically above R. A particle of mass m is attached at Q and hangs at rest in equilibrium. Find the distance of the particle below P.

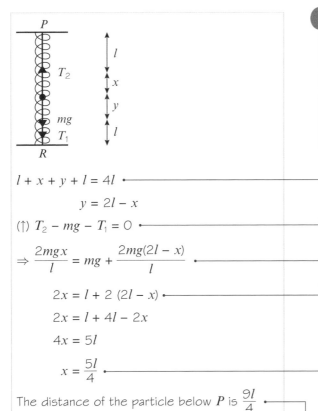

Draw a diagram showing the forces acting on the particle. Note that we have assumed that the lower spring is **stretched** and is therefore in **tension**. If the extension of the lower spring turns out to be negative, then it means the lower spring is in compression.

$l + x + y + l = 4l$ •——————————— Since P is $4l$ above R.

$\quad\quad\quad y = 2l - x$

$(\uparrow)\ T_2 - mg - T_1 = 0$ •——————— Since the mass is in equilibrium.

$\Rightarrow \dfrac{2mgx}{l} = mg + \dfrac{2mg(2l - x)}{l}$ •——— Use Hooke's law.

$\quad\quad 2x = l + 2(2l - x)$ •——————— Divide both sides by mg and multiply through by l.

$\quad\quad 2x = l + 4l - 2x$

$\quad\quad 4x = 5l$

$\quad\quad\quad x = \dfrac{5l}{4}$ •

Solve for x. The extension x is positive and so the top spring is in tension. The value of y is also positive. This demonstrates that the bottom spring is also in tension.

The distance of the particle below P is $\dfrac{9l}{4}$ •

Add on the natural length, l, of the spring.

Example ⑦   **SKILLS** CRITICAL THINKING

One end, A, of a light elastic string AB, of natural length 0.6 m and modulus of elasticity 10 N, is attached to a point on a fixed rough **plane**. The plane is **inclined** at an angle θ to the horizontal, where $\sin\theta = \dfrac{4}{5}$. A ball of mass 3 kg is attached to the end, B, of the string. The **coefficient of friction**, μ, between the ball and the plane is $\dfrac{1}{3}$. The ball rests in **limiting equilibrium**, on the point of sliding down the plane, with AB along the line of greatest **slope**.

a Find

  i the tension in the string

  ii the length of the string.

b If $\mu > \dfrac{1}{3}$, without doing any further calculation, state how your answer to part **a ii** would change.

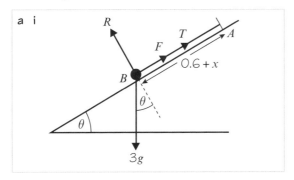

Draw a clear diagram showing all the forces. The ball is on the point of sliding **down** the plane, so the frictional force acts up the plane.

Let extension of string be x m.

$\sin \theta = \dfrac{4}{5}$ so $\cos \theta = \dfrac{3}{5}$

($\searcharrow$) $R = 3g \cos \theta = \dfrac{9g}{5}$

$F = \mu R = \dfrac{3g}{5}$

($\nearrow$) $T + F = 3g \sin \theta$

$T = 3g \sin \theta - F$

$T = \left(3g \times \dfrac{4}{5}\right) - \dfrac{3g}{5} = \dfrac{9g}{5} = 17.6 \, \text{N (3 s.f.)}$

ii $T = \dfrac{\lambda x}{l}$

$17.6 = \dfrac{10x}{0.6}$

so $x = 1.06 \, \text{m (3 s.f.)}$

Length of string $= 0.6 + 1.06$

$\qquad\qquad\qquad = 1.66 \, \text{m (3.s.f.)}$

b   If $\mu > \dfrac{1}{3}$ then

F would be greater as $F = \mu R$

T would be less as $T = 3g \sin \theta - F$

x would be less as $T = \dfrac{\lambda x}{l}$

so answer to part **a ii** would be less than 1.66 m

Resolving forces perpendicular to the plane.

Ball is in limiting equilibrium so $F = \mu R$

Resolving forces up the plane and substituting for T and F.

Watch out x is the extension in the string, so add the natural length to find the total length.

Problem-solving

The **coefficient of friction** is greater, which means that there is a greater force due to **friction** acting up the plane. The string has to produce less force to keep the ball in equilibrium, so less extension is required.

Exercise 2A   **SKILLS** ▷ PROBLEM-SOLVING

1. One end of a light elastic string is attached to a fixed point. A force of 4 N is applied to the other end of the string so as to stretch it. The natural length of the string is 3 m and the modulus of elasticity is λ N. Find the total length of the string when

   a $\lambda = 30$ **b** $\lambda = 12$ **c** $\lambda = 16$

2. The length of an elastic spring is reduced to 0.8 m when a force of 20 N compresses it. Given that the modulus of elasticity of the spring is 25 N, find its natural length.

3. An elastic spring of modulus of elasticity 20 N has one end fixed. When a particle of mass 1 kg is attached to the other end and hangs at rest, the total length of the spring is 1.4 m. The particle of mass 1 kg is removed and replaced by a particle of mass 0.8 kg. Find the new length of the spring.

4. A light elastic spring, of natural length a and modulus of elasticity λ, has one end fixed. A scale pan of mass M is attached to its other end and hangs in equilibrium. A mass m is gently placed in the scale pan. Find the distance of the new equilibrium position below the old one.

(P) 5 An elastic string has length a_1 when supporting a mass m_1 and length a_2 when supporting a mass m_2. Find the natural length and modulus of elasticity of the string.

(P) 6 When a weight, W N, is attached to a light elastic string of natural length l m the extension of the string is 10 cm. When W is increased by 50 N, the extension of the string is increased by 15 cm. Find W.

(E/P) 7 An elastic spring has natural length $2a$ and modulus of elasticity $2mg$. A particle of mass m is attached to the midpoint of the spring. One end of the spring, A, is attached to the floor of a room of height $5a$ and the other end is attached to the ceiling of the room at a point B vertically above A. The spring is modelled as light.
   a Find the distance of the particle below the ceiling when it is in equilibrium.   **(8 marks)**
   b In reality the spring may not be light. What effect will the model have had on the calculation of the distance of the particle below the ceiling?   **(1 mark)**

(E/P) 8 A **uniform** rod PQ, of mass 5 kg and length 3 m, has one end, P, smoothly hinged to a fixed point. The other end, Q, is attached to one end of a light elastic string of modulus of elasticity 30 N. The other end of the string is attached to a fixed point R which is on the same horizontal level as P with $RP = 5$ m. The system is in equilibrium and $\angle PQR = 90°$. Find
   a the tension in the string   **(5 marks)**
   b the natural length of the string.   **(3 marks)**

> **Problem-solving**
>
> First take **moments** about P.
>
> ← Mechanics 1 Section 8.1

(E/P) 9 A light elastic string AB has natural length l and modulus of elasticity $2mg$. Another light elastic string CD has natural length l and modulus of elasticity $4mg$. The strings are joined at their ends B and C and the end A is attached to a fixed point. A particle of mass m is hung from the end D and is at rest in equilibrium. Find the length AD.   **(7 marks)**

(E/P) 10 An elastic string PA has natural length 0.5 m and modulus of elasticity 9.8 N. The string PB is **inextensible**. The end A of the elastic string and the end B of the inextensible string are attached to two fixed points which are on the same horizontal level. The end P of each string is attached to a 2 kg particle. The particle hangs in equilibrium below AB, with PA making an angle of 30° with AB and PA perpendicular to PB. Find
   a the length of PA **(7 marks)**
   b the length of PB **(2 marks)**
   c the tension in PB.   **(2 marks)**

(E/P) 11 A particle of mass 2 kg is attached to one end P of a light elastic string PQ of modulus of elasticity 20 N and natural length 0.8 m. The end Q of the string is attached to a point on a rough plane which is inclined at an angle α to the horizontal, where $\tan \alpha = \dfrac{3}{4}$. The coefficient of friction between the particle and the plane is $\dfrac{1}{2}$. The particle rests in limiting equilibrium, on the point of sliding down the plane, with PQ along a line of greatest slope. Find
   a the tension in the string   **(5 marks)**
   b the length of the string.   **(2 marks)**

2.2 Hooke's law and dynamics problems

You can use Hooke's law to solve dynamics problems involving elastic strings or springs.

Example 8   **SKILLS**   CRITICAL THINKING

One end of a light elastic string, of natural length 0.5 m and modulus of elasticity 20 N, is attached to a fixed point A. The other end of the string is attached to a particle of mass 2 kg. The particle is held at a point which is 1.5 m below A and **released** from rest. Find:

a the initial acceleration of the particle

b the length of the string when the particle reaches its maximum speed.

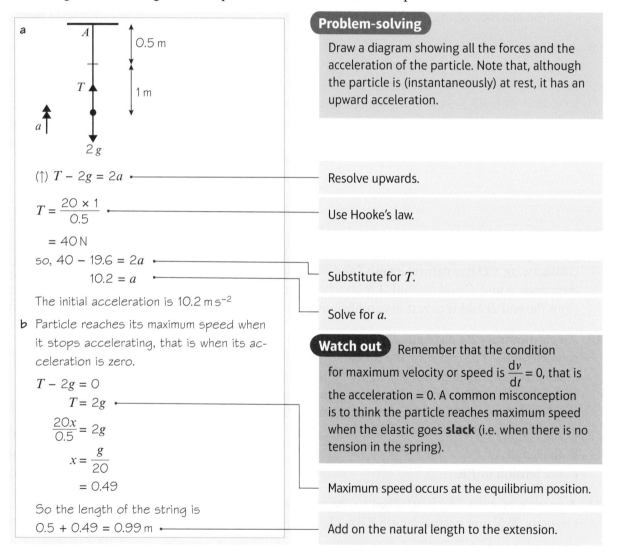

a

$(\uparrow)\ T - 2g = 2a$ ——— Resolve upwards.

$T = \dfrac{20 \times 1}{0.5}$ ——— Use Hooke's law.

$= 40\,\text{N}$

so, $40 - 19.6 = 2a$ ——— Substitute for T.

$10.2 = a$ ——— Solve for a.

The initial acceleration is $10.2\,\text{m s}^{-2}$

b Particle reaches its maximum speed when it stops accelerating, that is when its acceleration is zero.

$T - 2g = 0$

$\quad T = 2g$ ———

$\dfrac{20x}{0.5} = 2g$

$\quad x = \dfrac{g}{20}$

$= 0.49$

So the length of the string is

$0.5 + 0.49 = 0.99\,\text{m}$ ——— Add on the natural length to the extension.

Problem-solving

Draw a diagram showing all the forces and the acceleration of the particle. Note that, although the particle is (instantaneously) at rest, it has an upward acceleration.

Watch out Remember that the condition for maximum velocity or speed is $\dfrac{dv}{dt} = 0$, that is the acceleration = 0. A common misconception is to think the particle reaches maximum speed when the elastic goes **slack** (i.e. when there is no tension in the spring).

Maximum speed occurs at the equilibrium position.

Example **9**   **SKILLS** INTERPRETATION

A particle of mass 0.5 kg is attached to one end of a light elastic spring of natural length 1.5 m and modulus of elasticity 19.6 N. The other end of the spring is attached to a fixed point O on a rough plane which is inclined to the horizontal at an angle α, where $\tan\alpha = \dfrac{3}{4}$. The coefficient of friction between the particle and the plane is 0.2. The particle is held at rest on the plane at a point which is 1 m from O down a line of greatest slope of the plane. The particle is released from rest and moves down the slope. Find its initial acceleration.

Online Explore Hooke's law in dynamics problems using GeoGebra.

Problem-solving

Draw a diagram showing all four forces acting on the particle and the acceleration. Note that, since the spring is compressed, it produces a thrust, T, which acts **down** the plane. You can still apply Hooke's law in this situation.

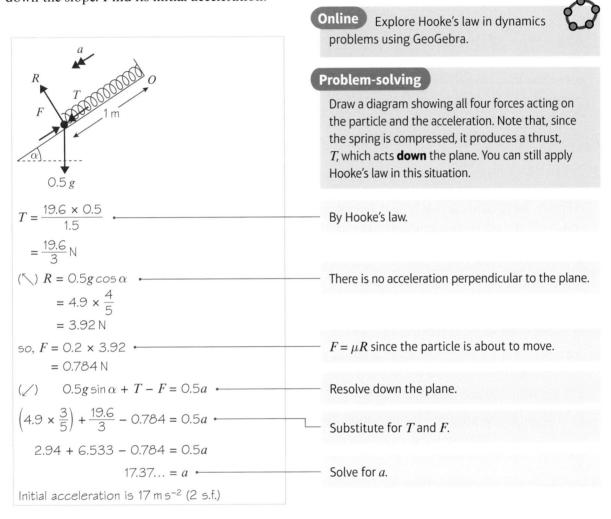

$T = \dfrac{19.6 \times 0.5}{1.5}$ ———————— By Hooke's law.

$\quad = \dfrac{19.6}{3}$ N

$(\nwarrow)\ R = 0.5g\cos\alpha$ ———————— There is no acceleration perpendicular to the plane.

$\quad\quad = 4.9 \times \dfrac{4}{5}$

$\quad\quad = 3.92$ N

so, $F = 0.2 \times 3.92$ ———————— $F = \mu R$ since the particle is about to move.

$\quad\quad = 0.784$ N

$(\nearrow)\quad 0.5g\sin\alpha + T - F = 0.5a$ ———————— Resolve down the plane.

$\left(4.9 \times \dfrac{3}{5}\right) + \dfrac{19.6}{3} - 0.784 = 0.5a$ ———————— Substitute for T and F.

$\quad 2.94 + 6.533 - 0.784 = 0.5a$

$\quad\quad\quad\quad\quad 17.37... = a$ ———————— Solve for a.

Initial acceleration is 17 m s^{-2} (2 s.f.)

Exercise **2B**   **SKILLS** INTERPRETATION

1 A particle of mass 4 kg is attached to one end P of a light elastic spring PQ, of natural length 0.5 m and modulus of elasticity 40 N. The spring rests on a **smooth** horizontal plane with the end Q fixed. The particle is held at rest and then released. Find the initial acceleration of the particle

  a if $PQ = 0.8$ m initially

  b if $PQ = 0.4$ m initially.

(E) **2** A particle of mass 0.4 kg is fixed to one end A of a light elastic spring AB, of natural length 0.8 m and modulus of elasticity 20 N. The other end B of the spring is attached to a fixed point. The particle hangs in equilibrium. It is then pulled vertically downwards through a distance 0.2 m and released from rest. Find the initial acceleration of the particle.   **(4 marks)**

(E/P) **3**

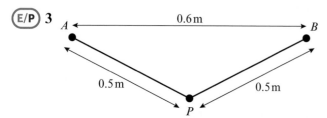

A particle P of mass 2 kg is attached to the midpoint of a light elastic string, of natural length 0.4 m and modulus of elasticity 20 N. The ends of the elastic string are attached to two fixed points A and B which are on the same horizontal level, with $AB = 0.6$ m. The particle is held in the position shown, with $AP = BP = 0.5$ m, and released from rest. Find the initial acceleration of the particle and state its direction.   **(5 marks)**

(E/P) **4** A particle of mass 2 kg is attached to one end P of a light elastic spring. The other end Q of the spring is attached to a fixed point O. The spring has natural length 1.5 m and modulus of elasticity 40 N. The particle is held at a point which is 1 m vertically above O and released from rest. Find the initial acceleration of the particle, stating its magnitude and direction.   **(5 marks)**

(E/P) **5** A particle of mass 1 kg is attached to one end of a light elastic spring of natural length 1.6 m and modulus of elasticity 21.5 N. The other end of the spring is attached to a fixed point O on a rough plane which is inclined to the horizontal at an angle α where $\tan \alpha = \dfrac{5}{12}$. The coefficient of friction between the particle and the plane is $\dfrac{1}{2}$. The particle is held at rest on the plane at a point which is 1.2 m from O down a line of greatest slope of the plane. The particle is released from rest and moves down the slope.

  a Find its initial acceleration.   **(6 marks)**

  b Without doing any further calculation, state how your answer to part **a** would change if the coefficient of friction between the particle and the plane was greater than $\dfrac{1}{2}$.   **(1 mark)**

> **Challenge**
>
> Two light elastic strings each have natural length l and modulus of elasticity λ.
> A particle P of mass 3 kg is attached to one end of each string. The other ends of the strings are attached to fixed points A and B, where AB is horizontal and $AB = 2x$ m.
> The particle is held at a point x m below the midpoint of AB and released from rest.
> The initial acceleration of the particle is $\dfrac{g}{2}$ m s^{-2}.
>
> **a** Show that when the particle is released the tension, T, in each string is $\dfrac{3\sqrt{2}g}{4}$ N.
>
> **b** Given that at the point the particle is released, each string has extended by $\dfrac{1}{4}$ of its natural length, find the modulus of elasticity for each string.

2.3 Elastic energy

You can find the energy stored in an elastic string or spring.

You can draw a **force–distance** diagram to show the extension x in an elastic string as a gradually increasing force is applied. The area under the force–distance graph is the **work done** in stretching the elastic string.

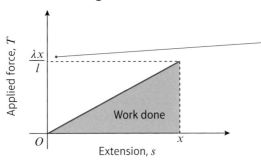

λ is the modulus of elasticity of the string and l is its natural length.

The applied force is always equal and opposite to the tension in the elastic string, T. This value increases as the string stretches.

Using the formula for the area of a triangle:

$$\text{Area} = \tfrac{1}{2} x \left(\frac{\lambda x}{l} \right)$$

$$= \frac{\lambda x^2}{2l}$$

Using integration:

$$\text{Area} = \int_0^x T \, \mathrm{d}s$$

$$= \int_0^x \frac{\lambda s}{l} \, \mathrm{d}s$$

$$= \left[\frac{\lambda s^2}{2l} \right]_0^x$$

$$= \frac{\lambda x^2}{2l}$$

- The work done in stretching an elastic string or spring of modulus of elasticity λ from its natural length l to a length $(l + x)$ is $\dfrac{\lambda x^2}{2l}$.

Watch out This rule applies as long as the extension is within the **elastic limit** of the string or spring.

When λ is measured in newtons and x and l are measured in metres, the work done is in **joules (J)**.

When a stretched string is released it will 'ping' back (i.e. return) to its natural length. In its stretched position it has the potential to do work, or elastic **potential energy** (this is also called **elastic energy**).

- The elastic potential **energy** (E.P.E.) stored in a stretched string or spring is exactly equal to the amount of work done to stretch the string or spring.

Links This is an application of the work–energy principle.
← Mechanics 2 Section 4.3

- The E.P.E. stored in an elastic string or spring of modulus of elasticity λ which has been stretched from its natural length l to a length $(l + x)$ is $\dfrac{\lambda x^2}{2l}$.

You can apply the same formulae for work done and elastic potential energy when an elastic string or spring is compressed.

Example **10** **SKILLS** PROBLEM-SOLVING

An elastic string has natural length 1.4 m and modulus of elasticity 6 N. Find the energy stored in the string when its length is 1.6 m.

Energy stored $= \dfrac{6 \times 0.2^2}{2 \times 1.4}$ ⟶ Use $\dfrac{\lambda x^2}{2l}$ with $x = 1.6 - 1.4 = 0.2$

$\qquad\qquad = 0.0857\,\text{J}$ (3 s.f.)

Example **11** **SKILLS** PROBLEM-SOLVING

A light elastic spring has natural length 0.6 m and modulus of elasticity 10 N. Find the work done in compressing the spring from a length of 0.5 m to a length of 0.3 m.

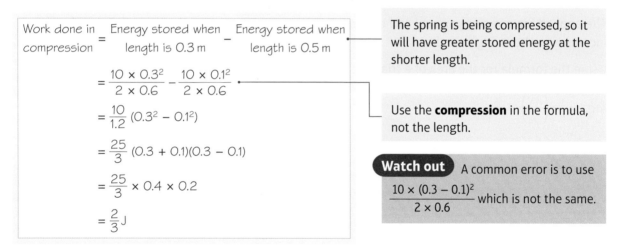

$$\begin{aligned}\text{Work done in compression} &= \begin{array}{c}\text{Energy stored when}\\ \text{length is } 0.3\,\text{m}\end{array} - \begin{array}{c}\text{Energy stored when}\\ \text{length is } 0.5\,\text{m}\end{array}\\[4pt] &= \frac{10 \times 0.3^2}{2 \times 0.6} - \frac{10 \times 0.1^2}{2 \times 0.6}\\[6pt] &= \frac{10}{1.2}(0.3^2 - 0.1^2)\\[6pt] &= \frac{25}{3}(0.3 + 0.1)(0.3 - 0.1)\\[6pt] &= \frac{25}{3} \times 0.4 \times 0.2\\[6pt] &= \frac{2}{3}\,\text{J}\end{aligned}$$

The spring is being compressed, so it will have greater stored energy at the shorter length.

Use the **compression** in the formula, not the length.

Watch out A common error is to use $\dfrac{10 \times (0.3 - 0.1)^2}{2 \times 0.6}$ which is not the same.

Exercise **2C** **SKILLS** PROBLEM-SOLVING

1 An elastic spring has natural length 0.6 m and modulus of elasticity 8 N. Find the work done when the spring is stretched from its natural length to a length of 1 m.

2 An elastic spring, of natural length 0.8 m and modulus of elasticity 4 N, is compressed to a length of 0.6 m. Find the elastic potential energy stored in the spring.

3 An elastic string has natural length 1.2 m and modulus of elasticity 10 N. Find the work done when the string is stretched from a length of 1.5 m to a length 1.8 m.

4 An elastic spring has natural length 0.7 m and modulus of elasticity 20 N. Find the work done when the spring is stretched from a length

 a 0.7 m to 0.9 m

 b 0.8 m to 1.0 m

 c 1.2 m to 1.4 m

Hint Note that your answers to **a**, **b** and **c** are all different.

(E) **5** A light elastic spring has natural length 1.2 m and modulus of elasticity 10 N. One end of the spring is attached to a fixed point. A particle of mass 2 kg is attached to the other end and hangs in equilibrium. Find the energy stored in the spring. **(3 marks)**

(E/P) **6** An elastic string has natural length a. One end is fixed. A particle of mass $2m$ is attached to the free end and hangs in equilibrium, with the length of the string $3a$. Find the elastic potential energy stored in the string. **(3 marks)**

(E/P) **7**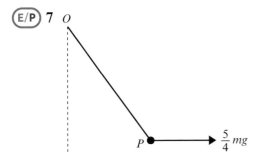

A particle P of mass m is attached to one end of a light elastic string, of natural length a and modulus of elasticity $2mg$. The other end of the string is attached to a fixed point O.

The particle P is held in equilibrium by a horizontal force of magnitude $\frac{5}{4}mg$ applied to P.

This force acts in the vertical plane containing the string, as shown in the diagram.

Find

a the tension in the string **(5 marks)**

b the elastic energy stored in the string. **(4 marks)**

2.4 Problems involving elastic energy

You can solve problems involving elastic energy using the principle of conservation of mechanical energy and the work–energy principle.

■ **When no external forces (other than gravity) act on a particle, then the sum of its kinetic energy, gravitational potential energy and elastic potential energy remains constant.**

> **Links** This is an application of the principle of conservation of mechanical energy.
> ← Mechanics 2 Section 4.3

If a particle which is attached to an elastic spring or string is subject to a **resistance** as it moves, you will need to apply the work–energy principle.

Example **12** **SKILLS** INTERPRETATION

A light elastic string, of natural length 1.6 m and modulus of elasticity 10 N, has one end fixed at a point A on a smooth horizontal table. A particle of mass 2 kg is attached to the other end of the string. The particle is held at the point A and **projected** horizontally along the table with speed 2 m s⁻¹. Find how far it travels before first coming to instantaneous rest.

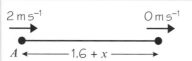

Suppose that the extension of the string
when the particle comes to rest is x.

K.E. lost by the particle = E.P.E. gained by the
                                     string

$$\frac{1}{2}mv^2 = \frac{\lambda x^2}{2l}$$

$$\frac{1}{2} \times 2 \times 2^2 = \frac{10x^2}{2 \times 1.6}$$

$$1.28 = x^2$$

$$1.13\ldots = x$$

Total distance travelled is 2.73 m (3 s.f.)

	Draw a simple diagram showing the initial and final positions of the particle.
	You can apply the **conservation of energy** principle since the table is smooth.
	Note that you do not need to consider the energy of the particle in any intermediate positions.
	Add on the natural length of the string.

It is important to realise that in the example above, the particle is not in equilibrium when it comes to instantaneous rest. Therefore you cannot use forces to solve this type of problem. The particle has an acceleration and will immediately 'spring' back towards A.

Example **13**   **SKILLS**   PROBLEM-SOLVING

A particle of mass 0.5 kg is attached to one end of an elastic string, of natural length 2 m and modulus of elasticity 19.6 N. The other end of the elastic string is attached to a point O. In fact, the particle is released from the point O. Find the greatest distance it will reach below O.

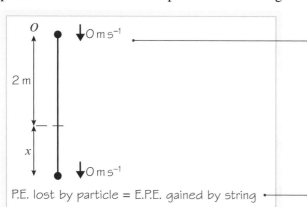

Draw a diagram showing the initial and final positions of the particle. Let the extension of the string be x when the particle comes to rest.

P.E. lost by particle = E.P.E. gained by string

$$mgh = \frac{\lambda x^2}{2l}$$

$$0.5g(2 + x) = \frac{19.6x^2}{4}$$

$$2 + x = x^2$$

$$0 = x^2 - x - 2$$

$$0 = (x - 2)(x + 1)$$

$$x = 2 \text{ or } -1$$

Hence greatest distance reached below O is
4 m.

There is no K.E. involved as the particle starts at rest and finishes at rest. Assuming no air resistance, energy will be conserved.

Problem-solving

x is the extension in the string so it must be positive. Ignore the negative solution, and remember to add the natural length of the string.

Example (14)　　**SKILLS**　INTERPRETATION

A light elastic spring, of natural length 1 m and modulus of elasticity 10 N, has one end attached to a fixed point A. A particle of mass 2 kg is attached to the other end of the spring and is held at a point B which is 0.8 m vertically below A. The particle is projected vertically downwards from B with speed 2 m s^{-1}. Find the distance it travels before first coming to rest.

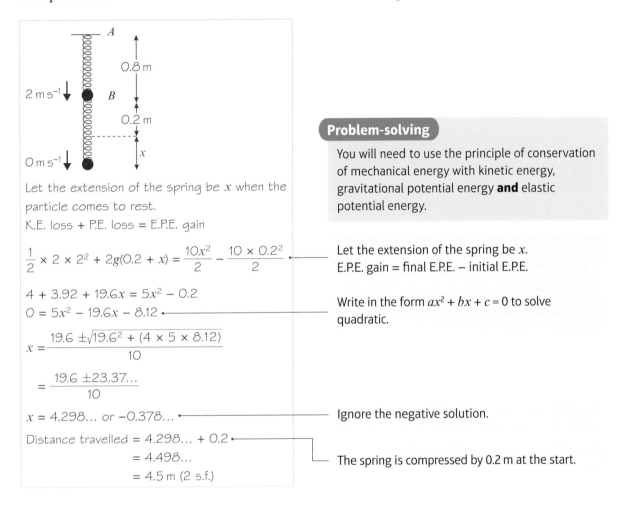

Let the extension of the spring be x when the particle comes to rest.

K.E. loss + P.E. loss = E.P.E. gain

$$\frac{1}{2} \times 2 \times 2^2 + 2g(0.2 + x) = \frac{10x^2}{2} - \frac{10 \times 0.2^2}{2}$$

$$4 + 3.92 + 19.6x = 5x^2 - 0.2$$

$$0 = 5x^2 - 19.6x - 8.12$$

$$x = \frac{19.6 \pm \sqrt{19.6^2 + (4 \times 5 \times 8.12)}}{10}$$

$$= \frac{19.6 \pm 23.37...}{10}$$

$x = 4.298...$ or $-0.378...$

Distance travelled = 4.298... + 0.2

　　　　　　　　　= 4.498...

　　　　　　　　　= 4.5 m (2 s.f.)

Problem-solving

You will need to use the principle of conservation of mechanical energy with kinetic energy, gravitational potential energy **and** elastic potential energy.

Let the extension of the spring be x.
E.P.E. gain = final E.P.E. − initial E.P.E.

Write in the form $ax^2 + bx + c = 0$ to solve quadratic.

Ignore the negative solution.

The spring is compressed by 0.2 m at the start.

Example (15)　　**SKILLS**　CRITICAL THINKING

A light elastic spring, of natural length 0.5 m and modulus of elasticity 10 N, has one end attached to a point A on a rough horizontal plane. The other end is attached to a particle P of mass 0.8 kg. The coefficient of friction between the particle and the plane is 0.4. Initially the particle lies on the plane with $AP = 0.5$ m. It is then projected with speed 2 m s^{-1} away from A along the plane. Find the distance moved by P before it first comes to rest.

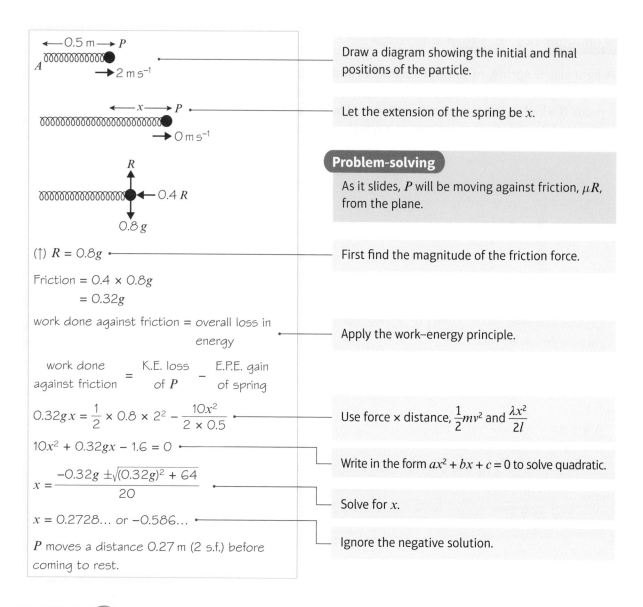

Draw a diagram showing the initial and final positions of the particle.

Let the extension of the spring be x.

Problem-solving

As it slides, P will be moving against friction, μR, from the plane.

$(\uparrow)$ $R = 0.8g$ — First find the magnitude of the friction force.

Friction $= 0.4 \times 0.8g$

$\qquad = 0.32g$

work done against friction = overall loss in energy — Apply the work–energy principle.

$$\begin{array}{c} \text{work done} \\ \text{against friction} \end{array} = \begin{array}{c} \text{K.E. loss} \\ \text{of } P \end{array} - \begin{array}{c} \text{E.P.E. gain} \\ \text{of spring} \end{array}$$

$0.32gx = \dfrac{1}{2} \times 0.8 \times 2^2 - \dfrac{10x^2}{2 \times 0.5}$ — Use force × distance, $\dfrac{1}{2}mv^2$ and $\dfrac{\lambda x^2}{2l}$

$10x^2 + 0.32gx - 1.6 = 0$ — Write in the form $ax^2 + bx + c = 0$ to solve quadratic.

$x = \dfrac{-0.32g \pm \sqrt{(0.32g)^2 + 64}}{20}$ — Solve for x.

$x = 0.2728\ldots$ or $-0.586\ldots$ — Ignore the negative solution.

P moves a distance 0.27 m (2 s.f.) before coming to rest.

Exercise **2D** **SKILLS** **INTERPRETATION**

P **1** An elastic string, of natural length l and modulus of elasticity mg, has one end fixed to a point A on a smooth horizontal table. The other end is attached to a particle P of mass m. The particle is held at a point on the table with $AP = \dfrac{3}{2}l$ and is released. Find the speed of the particle when the string reaches its natural length.

P **2** A particle of mass m is **suspended** from a fixed point O by a light elastic string, of natural length a and modulus of elasticity $4mg$. The particle is pulled vertically downwards a distance d from its equilibrium position and released from rest. If the particle just reaches O, find d.

E/P **3** A light elastic spring of natural length $2l$ has its ends attached to two points P and Q which are at the same horizontal level. The length PQ is $2l$. A particle of mass m is fastened to the midpoint of the spring and is held at the midpoint of PQ. The particle is released from rest and first comes to instantaneous rest when both parts of the spring make an angle of $60°$ with the line PQ.

    a Find the modulus of elasticity of the spring. **(6 marks)**

    b Suggest one way in which the model could be refined to make it more realistic. **(1 mark)**

(E/P) 4 A light elastic string, of natural length 1 m and modulus of elasticity 21.6 N, has one end attached to a fixed point O. A particle of mass 2 kg is attached to the other end. The particle is held at a point which is 3 m vertically below O and released from rest. Find

    a the speed of the particle when the string first becomes slack **(5 marks)**

    b the distance from O when the particle first comes to rest. **(3 marks)**

(E/P) 5 A particle P is attached to one end of a light elastic string of natural length a. The other end of the string is attached to a fixed point O. When P hangs at rest in equilibrium, the distance OP is $\frac{5a}{3}$. The particle is now projected vertically downwards from O with speed U and first comes to instantaneous rest at a distance $\frac{10a}{3}$ below O. Find U in terms of a and g. **(7 marks)**

(E/P) 6 A particle P of mass 1 kg is attached to the midpoint of a light elastic string, of natural length 3 m and modulus λ N. The ends of the string are attached to two points A and B on the same horizontal level with $AB = 3$ m. The particle is held at the midpoint of AB and released from rest. The particle falls vertically and comes to instantaneous rest at a point which is 1 m below the midpoint of AB. Find

    a the value of λ **(5 marks)**

    b the speed of P when it is 0.5 m below the initial position. **(5 marks)**

(E/P) 7 A light elastic string of natural length 2 m and modulus of elasticity 117.6 N has one end attached to a fixed point O. A particle P of mass 3 kg is attached to the other end. The particle is held at O and released from rest.

    a Find the distance fallen by P before it first comes to rest. **(5 marks)**

    b Find the greatest speed of P during the fall. **(4 marks)**

Problem-solving

P will be travelling at its greatest speed when the acceleration is zero.

(E/P) 8 A particle P of mass 2 kg is attached to one end of a light elastic string of natural length 1 m and modulus of elasticity 40 N. The other end of the string is fixed to a point O on a rough plane which is inclined at an angle α, where $\tan \alpha = \frac{3}{4}$. The particle is held at O and released from rest. Given that P comes to rest after moving 2 m down the plane, find the coefficient of friction between the particle and the plane. **(4 marks)**

Challenge

An elastic string of natural length l m is suspended from a fixed point O.
When a mass of M kg is attached to the other end of the string, its extension
is $\frac{l}{10}$ m. An additional M kg is then attached to the end of the string.
Show that the work done in producing the additional extension is $\frac{3Mgl}{20}$ J.

Chapter review **2** **SKILLS** PROBLEM-SOLVING

P **1** A particle of mass m is supported by two light elastic strings, each of natural length a and modulus of elasticity $\dfrac{15mg}{16}$. The other ends of the strings are attached to two fixed points A and B where A and B are in the same horizontal line with $AB = 2a$. When the particle hangs at rest in equilibrium below AB, each string makes an angle θ with the vertical.

  a Verify that $\cos\theta = \dfrac{4}{5}$.

  b How much work must be done to raise the particle to the midpoint of AB?

2 A light elastic spring is such that a weight of magnitude W resting on the spring produces a compression a. The weight W is allowed to fall onto the spring from a height of $\dfrac{3a}{2}$ above it. Find the maximum compression of the spring in the subsequent motion.

3 A light elastic string of natural length $0.5\,\text{m}$ is stretched between two points P and Q on a smooth horizontal table. The distance PQ is $0.75\,\text{m}$ and the tension in the string is $15\,\text{N}$.

  a Find the modulus of elasticity of the string.

  A particle of mass $0.5\,\text{kg}$ is attached to the midpoint of the string. The particle is pulled $0.1\,\text{m}$ towards Q and released from rest.

  b Find the speed of the particle as it passes through the midpoint of PQ.

P **4** A particle of mass m is attached to two strings AP and BP. The points A and B are on the same horizontal level and $AB = \dfrac{5a}{4}$.

  The string AP is inextensible and $AP = \dfrac{3a}{4}$.

  The string BP is elastic and $BP = a$.

  The modulus of elasticity of BP is λ. Show that the natural length of BP is $\dfrac{5\lambda a}{3mg + 5\lambda}$

P **5** A light elastic string, of natural length a and modulus of elasticity $5mg$, has one end attached to the base of a vertical wall. The other end of the string is attached to a small ball of mass m.

  The ball is held at a distance $\dfrac{3a}{2}$ from the wall, on a rough horizontal plane, and released from rest.

  The coefficient of friction between the ball and the plane is $\dfrac{1}{5}$.

  a Find, in terms of a and g, the speed V of the ball as it hits the wall.

  The ball rebounds (i.e. bounces back) from the wall with speed $\dfrac{2V}{5}$. The string stays slack.

  b Find the distance from the wall at which the ball comes to rest.

(E/P) **6** A light elastic string has natural length l and modulus $2mg$. One end of the string is attached to a particle P of mass m. The other end is attached to a fixed point C on a rough horizontal plane. Initially P is at rest at a point D on the plane where $CD = \dfrac{4l}{3}$

a Given that P is in limiting equilibrium, find the coefficient of friction between P and the plane.
(5 marks)

The particle P is now moved away from C to a point E on the plane where $CE = 2l$

b Find the speed of P when the string returns to its natural length. **(5 marks)**

c Find the total distance moved by P before it comes to rest. **(4 marks)**

(E/P) **7** A light elastic string of natural length 0.2 m has its ends attached to two fixed points A and B which are on the same horizontal level with $AB = 0.2$ m. A particle of mass 5 kg is attached to the string at the point P where $AP = 0.15$ m. The system is released and P hangs in equilibrium below AB with $\angle APB = 90°$.

a If $\angle BAP = \theta$, show that the ratio of the extension of AP and BP is

$$\frac{4\cos\theta - 3}{4\sin\theta - 1}$$
(4 marks)

b Hence show that

$$\cos\theta\,(4\cos\theta - 3) = 3\sin\theta\,(4\sin\theta - 1)$$
(4 marks)

(E/P) **8** A particle of mass 3 kg is attached to one end of a light elastic string, of natural length 1 m and modulus of elasticity 14.7 N. The other end of the string is attached to a fixed point. The particle is held in equilibrium by a horizontal force of magnitude 9.8 N with the string inclined to the vertical at an angle θ.

a Find the value of θ. **(3 marks)**

b Find the extension of the string. **(3 marks)**

c If the horizontal force is removed, find the magnitude of the least force that will keep the string inclined at the same angle. **(2 marks)**

(E/P) **9** Two points A and B are on the same horizontal level with $AB = 3a$. A particle P of mass m is joined to A by a light inextensible string of length $4a$ and is joined to B by a light elastic string, of natural length a and modulus of elasticity $\dfrac{mg}{4}$. The particle P is held at a point C, such that $BC = a$ and both strings are **taut**. The particle P is released from rest.

a Show that when AP is vertical the speed of P is $2\sqrt{ga}$ **(6 marks)**

b Find the tension in the elastic string in this position. **(4 marks)**

Challenge

A bungee jumper attaches one end of an elastic rope to both ankles. The other end is attached to the platform on which he stands.

The bungee jumper is modelled as a particle of mass m kg attached to an elastic string of natural length l m with modulus of elasticity λ N.

a Show that the maximum distance the jumper **descends** after jumping off the platform is $l + k + \sqrt{2kl + k^2}$ where $k = \dfrac{mgl}{\lambda}$

b Suggest a **refinement** to this model that would result in
  i a greater maximum descent
  ii a smaller maximum descent.

Summary of key points

1 When an elastic string or spring is stretched, the tension, T, produced is proportional to the extension, x.

- $T \propto x$
- $T = kx$, **where k is a constant**

The constant k depends on the unstretched length of the spring or string, l, and the **modulus of elasticity** of the string or spring, λ.

- $T = \dfrac{\lambda x}{l}$

This relationship is called **Hooke's law**.

2 The area under a **force–distance** graph is the **work done** in stretching an elastic string or spring. The work done in stretching or compressing an elastic string or spring with modulus of elasticity λ from its natural length, l to a length $(l + x)$ is $\dfrac{\lambda x^2}{2l}$

When λ is measured in newtons and x and l are measured in metres, the work done is in **joules (J)**.

3 The **elastic potential energy** (E.P.E.) stored in a stretched string or spring is exactly equal to the amount of work done to stretch the string or spring.

The E.P.E. stored in a string or spring of modulus of elasticity λ which has been stretched from its natural length l, to a length $(l + x)$ is $\dfrac{\lambda x^2}{2l}$

4 When no external forces (other than gravity) act on a particle, then the sum of its kinetic energy, gravitational potential energy and elastic potential energy remains constant.

3 DYNAMICS

Learning objectives

After completing this chapter you should be able to:

● Use calculus to apply Newton's laws to a particle moving in a straight line
→ pages 41–46

● Use Newton's law of gravitation to solve problems involving a particle moving away from
(or towards) the Earth's surface
→ pages 46–49

● Solve problems involving a particle moving in a straight line with simple harmonic motion
→ pages 50–59

● Investigate the motion of a particle attached to an elastic spring or string and oscillating
in a horizontal line
→ pages 59–64

● Investigate the motion of a particle attached to an elastic spring or string and oscillating
in a vertical line
→ pages 64–76

Prior knowledge check

1 A car of mass 1500 kg moves in a straight line. The total resistance
to motion of the car is modelled as a constant force of magnitude
80 N. The car brakes with a constant force, bringing it to rest from
a speed of 30 m s^{-1} in a distance of 60 m.

a Find the magnitude of the braking force.

b Find the total work done in bringing the
car to rest.
← **Mechanics 2 Section 4.1**

2 A particle travels along the positive x-axis
with acceleration $a = \dfrac{e^{2t} - 3t^2}{5}$ at time t s.
The particle starts from rest at the origin, O. Find

a the velocity of the particle when $t = 3$ s

b the displacement from O when $t = 2$ s

← **Mechanics 2 Section 2.3**

3 An elastic string of natural length 2.5 m is fixed at one end and is
stretched to a length of 3.4 m by a force of 5 N. Find the modulus
of elasticity of the string.
← **Mechanics 3 Section 2.1**

These toys can be modelled
as particles hanging vertically
by an elastic string. When
stretched then released
they will experience simple
harmonic motion, and their
displacement–time graphs will
be in the shape of a sine curve.

3.1 Motion in a straight line with variable force

You can use calculus to apply Newton's second law $F = ma$ to a particle moving in a straight line when the applied force is variable.

The applied force F can be a function of the displacement x, or time t.

Links When the acceleration is a function of x, t or v, it can be written as $\dfrac{dv}{dt}$, $v\dfrac{dv}{dx}$ or $\dfrac{d^2x}{dt^2}$
← **Mechanics 3 Sections 1.1, 1.2**

Suppose F is a function of time, then using $a = \dfrac{dv}{dt}$

$$m\frac{dv}{dt} = F$$

$$\int m\,dv = \int F\,dt$$ —— Separate the variables.

$$mv = \int F\,dt$$ —— Integrate both sides. Mass is assumed to be constant.

Now suppose F is a function of displacement, then using $a = v\dfrac{dv}{dx}$

$$mv\frac{dv}{dx} = F$$

$$\int mv\,dv = \int F\,dx$$ —— Separate the variables.

So $\dfrac{1}{2}mv^2 = \int F\,dx$ —— Integrate both sides.

When you work out $\int F\,dt$ or $\int F\,dx$ you must remember to add a **constant of integration**. You will often be given boundary conditions that allow you to work out the value of this constant.

Example 1   SKILLS   PROBLEM-SOLVING

A particle P of mass $0.5\,\text{kg}$ is moving along the x-axis. At time t seconds the force acting on P has magnitude $(5t^2 + e^{0.2t})\,\text{N}$ and acts in the direction OP. When $t = 0$, P is at rest at O. Calculate

a the speed of P when $t = 2$

b the distance OP when $t = 3$

a $F = ma$

$5t^2 + e^{0.2t} = 0.5a$

$0.5\dfrac{dv}{dt} = 5t^2 + e^{0.2t}$

$0.5v = \dfrac{5}{3}t^3 + \dfrac{1}{0.2}e^{0.2t} + C$

$v = \dfrac{10}{3}t^3 + 10e^{0.2t} + D$

$t = 0,\ v = 0 \Rightarrow 0 = 0 + 10 + D$

$\therefore D = -10$ and $v = \dfrac{10}{3}t^3 + 10e^{0.2t} - 10$

Use $F = ma$ with $F = 5t^2 + e^{0.2t}$ and $m = 0.5$

Replace a with $\dfrac{dv}{dt}$

Integrate with respect to t. Remember to add the constant of integration.

Divide by 0.5. Change C to D instead of dividing.

Use $v = 0$ when $t = 0$ to find the value of D and complete the expression for v.

$t = 2, v = \dfrac{80}{3} + 10e^{0.4} - 10 = 31.58\ldots$

Now substitute $t = 2$ to obtain the required value.

When $t = 2$ the speed of P is $31.6\,\text{m s}^{-1}$ (3 s.f.).

b $v = \dfrac{10}{3}t^3 + 10e^{0.2t} - 10$

$\dfrac{dx}{dt} = \dfrac{10}{3}t^3 + 10e^{0.2t} - 10$

Replace v with $\dfrac{dx}{dt}$

$x = \dfrac{10}{12}t^4 + \dfrac{10}{0.2}e^{0.2t} - 10t + K$

Integrate with respect to t. Use a different letter for the constant of integration.

$t = 0, x = 0 \Rightarrow 0 = 0 + 50 - 0 + K$

$\therefore K = -50$ and $x = \dfrac{10}{12}t^4 + \dfrac{10}{0.2}e^{0.2t} - 10t - 50$

Use the initial conditions to find K.

$t = 3 \Rightarrow x = \dfrac{10}{12} \times 3^4 + \dfrac{10}{0.2}e^{0.6} - 10 \times 3 - 50 = 78.60\ldots$

Now substitute $t = 3$ to obtain the required value.

When $t = 3$, OP is $78.6\,\text{m}$ (3 s.f.).

Example **2**   **SKILLS**   INTERPRETATION

A rock of mass $0.2\,\text{kg}$ is moving on a smooth horizontal sheet of ice. At time t seconds (where $t \geqslant 0$) a horizontal force of magnitude $2t^2\,\text{N}$ and constant direction acts on the rock. When $t = 0$ the rock is moving in the same direction as the force and has speed $6\,\text{m s}^{-1}$. When $t = T$ the rock has speed $36\,\text{m s}^{-1}$. Calculate the value of T.

$F = ma$

$0.2\dfrac{dv}{dt} = 2t^2$

Use $F = ma$ with $F = 2t^2$ and $m = 0.2$

As F is a function of t replace a with $\dfrac{dv}{dt}$

$0.2v = \dfrac{2}{3}t^3 + C$

$t = 0, v = 6 \Rightarrow 0.2 \times 6 = 0 + C$

Integrate with respect to t and use the initial conditions to find the value of C.

$\therefore C = 1.2$ and $0.2v = \dfrac{2}{3}t^3 + 1.2$

$t = T, v = 36 \Rightarrow 0.2 \times 36 + \dfrac{2}{3}T^3 + 1.2$

Substitute $t = T$ and solve for T.

$T^3 = \dfrac{3}{2} \times 6$

$T = 2.080\ldots$

$T = 2.08$ (3 s.f.)

Suppose F is a function of displacement, x, then using $a = \dfrac{d}{dx}\left(\dfrac{1}{2}v^2\right) = v\dfrac{dv}{dx}$

$mv\dfrac{dv}{dx} = F$

$\int mv\,dv = \int F\,dx$ Separate the variables.

$\dfrac{1}{2}mv^2 = \int F\,dx + C$

where C is the constant of integration m is constant so $\int mv\,dv = \dfrac{1}{2}mv^2 + C$

Example (3) **SKILLS** PROBLEM-SOLVING

A particle P of mass 1.5 kg is moving in a straight line. The force acting on P has magnitude $(8 - 2\cos x)$ N, where x metres is the distance OP, and acts in the direction OP. When P passes through O its speed is 4 m s^{-1}. Calculate the speed of P when $x = 2$

$F = ma$	
$8 - 2\cos x = 1.5a$	Use $F = ma$ with $F = 8 - 2\cos x$ and $m = 1.5$
$1.5v\dfrac{dv}{dx} = 8 - 2\cos x$	As F is a function of x, replace a with $v\dfrac{dv}{dx}$
$1.5\int v\,dv = \int 8 - 2\cos x\,dx$	
$1.5 \times \dfrac{1}{2}v^2 = 8x - 2\sin x + C$	Separate the variables and integrate with respect to x.
$x = 0, v = 4 \Rightarrow 1.5 \times \dfrac{1}{2} \times 4^2 = 0 - 0 + C$	
$\therefore C = 12$ and $0.75v^2 = 8x - 2\sin x + 12$	Use the initial conditions to find C.
$x = 2 \Rightarrow 0.75v^2 = (8 \times 2) - 2\sin 2 + 12$	Substitute $x = 2$ to find the required speed.
$v = 5.908...$	Remember that as calculus has been used,
When $x = 2$ the speed of P is 5.91 m s^{-1} (3 s.f.).	x must be in radians.

Example (4) **SKILLS** CRITICAL THINKING

A stone S of mass 0.5 kg is moving in a straight line on a smooth horizontal floor. When S is a distance x metres from a fixed point on the line, A, a force of magnitude $(5 + 7\cos x)$ N acts on S in the direction AS. Given that S passes through A with speed 2 m s^{-1}, calculate

a the speed of S as it passes through the point B, where $x = 3$

b the work done by the force in moving S from A to B.

a $F = ma$	Use $F = ma$ with $F = 5 + 7\cos x$ and $m = 0.5$
$0.5v\dfrac{dv}{dx} = 5 + 7\cos x$	As F is a function of x, replace a with $v\dfrac{dv}{dx}$
$\int 0.5v\,dv = \int(5 + 7\cos x)\,dx$	
$0.5 \times \dfrac{1}{2}v^2 = 5x + 7\sin x + C$	Separate the variables and integrate. Use
$x = 0, v = 2 \Rightarrow 0.5 \times \dfrac{1}{2} \times 2^2 = C$	the initial conditions to find the value of C.
$\therefore C = 1$ and $\dfrac{1}{4}v^2 = 5x + 7\sin x + 1$	
$x = 3 \Rightarrow v^2 = 4(15 + 7\sin 3 + 1)$	
$v^2 = 67.95...$	
$v = 8.243...$	Substitute $x = 3$ to find the value of v.
S passes through B with speed 8.24 m s^{-1} (3 s.f.)	
b Work done = increase in K.E.	Use the work–energy principle. The work
$= \dfrac{1}{2} \times 0.5 \times 67.95 - \dfrac{1}{2} \times 0.5 \times 2^2$	done is equal to the increase in kinetic
$= 15.98...$	energy. ← **Mechanics 2 Section 4.3**
The work done is 16.0 J (3 s.f.).	

- In forming an equation of motion, forces that tend to decrease the displacement are negative and forces that tend to increase the displacement are positive.

Exercise 3A  **SKILLS** CRITICAL THINKING

1 A particle P of mass $0.2\,kg$ is moving on the x-axis. At time t seconds P is x metres from the origin O. The force acting on P has magnitude $2\cos t\,N$ and acts in the direction OP. When $t = 0$, P is at rest at O. Calculate
   a the speed of P when $t = 2$
   b the speed of P when $t = 3$
   c the time when P first comes to instantaneous rest
   d the distance OP when $t = 2$
   e the distance OP when P first comes to instantaneous rest.

2 A van of mass $1200\,kg$ moves along a horizontal straight road. At time t seconds, the **resultant force** acting on the car has magnitude $\dfrac{60\,000}{(t + 5)^2}\,N$ and acts in the direction of motion of the van. When $t = 0$, the van is at rest. The speed of the van approaches a limiting value $V\,m\,s^{-1}$. Find
   a the value of V
   b the distance moved by the van in the first 4 seconds of its motion.

3 A particle P of mass $0.8\,kg$ is moving along the x-axis. At time $t = 0$, P passes through the origin O, moving in the positive x direction. At time t seconds, $OP = x$ metres and the velocity of P is $v\,m\,s^{-1}$. The resultant force acting on P has magnitude $\dfrac{1}{6}(15 - x)\,N$, and acts in the positive x direction. The maximum speed of P is $12\,m\,s^{-1}$.
   a Explain why the maximum speed of P occurs when $x = 15$
   b Find the speed of P when $t = 0$

4 A particle P of mass $0.75\,kg$ is moving in a straight line. At time t seconds after it passes through a fixed point on the line, O, the distance OP is x metres and the force acting on P has magnitude $(2e^{-x} + 2)\,N$ and acts in the direction OP. Given that P passes through O with speed $5\,m\,s^{-1}$, calculate the speed of P when
   a $x = 3$
   b $x = 7$
   c Calculate the work done by the force in moving the particle from the point where $x = 3$ to the point where $x = 7$

5 A particle P of mass $0.5\,kg$ moves away from the origin O along the positive x-axis. When $OP = x$ metres the force acting on P has magnitude $\dfrac{3}{x + 2}\,N$ and is directed away from O. When $x = 0$ the speed of P is $1.5\,m\,s^{-1}$. Find the value of x when the speed of P is $2\,m\,s^{-1}$.

(E/P) **6** A particle P of mass 250 g moves along the x-axis in the positive direction. At time $t = 0$, P passes through the origin with speed $10\,\text{m s}^{-1}$. At time t seconds, the distance OP is x metres and the speed of P is $v\,\text{m s}^{-1}$. The resultant force acting on P is directed towards the origin and has magnitude $\dfrac{8}{(t + 1)^2}\,\text{N}$

a Show that $v = 2\left(\dfrac{16}{t + 1} - 11\right)$ **(5 marks)**

b Find the value of x when $t = 5$ **(5 marks)**

(E/P) **7** A particle P of mass 0.6 kg moves along the x-axis in the positive direction. A single force acting on P is directed towards the origin, O, and has magnitude $\dfrac{k}{(x + 2)^2}\,\text{N}$ where $OP = x$ metres and k is a constant.

> **Problem-solving**
>
> The boundary conditions give the velocity for two different displacements. You need to set up and solve two simultaneous equations.

At time $t = 0$, P passes through the origin. When $x = 3$ the speed of P is $5\,\text{m s}^{-1}$, when $x = 8$ the speed of P is $\sqrt{3}\,\text{m s}^{-1}$. Find the value of k. **(6 marks)**

> **Challenge**
>
> A particle P of mass m kg is acted on by a single force, F N, and moves in a straight line, passing a fixed point O at time $t = 0$. At the point when the displacement of the particle from O is x m, the force acts in the direction OP and has magnitude
> $$F = 3x^2 - \sqrt[3]{x}\,\text{N}$$
>
> **a** Show that the work done by the force between times $t = a$ and $t = b$ is independent of the velocity of the particle at the point when it passes O.
>
> **b** Find the work done by the force in the first 6 seconds of the motion of the particle.

3.2 Newton's law of gravitation

You can use Newton's law of **gravitation** to solve problems involving a particle moving away from (or towards) the Earth's surface.

Newton's law of gravitation states

■ The **force of attraction** between two bodies of masses M_1 and M_2 is **directly proportional** to the product of their masses and **inversely proportional** to the square of the distance between them.

This law is sometimes referred to as the inverse square law. It can be expressed mathematically by the following equation

$$F = \frac{GM_1M_2}{d^2}$$

where G is a constant known as the constant of gravitation.

> **Watch out** Newton's law of gravitation should be used when modelling large changes in distance relative to the sizes of the bodies, such as a rocket being launched into orbit. For small changes in height (such as when a ball is thrown into the air), gravity can be modelled as a constant force.

This force causes particles (and bodies) to fall to the Earth and the Moon to orbit the Earth.

Relationship between *G* and *g*

When a particle of mass m is resting on the surface of the Earth, the force with which the Earth attracts the particle has magnitude mg and is directed towards the centre of the Earth.

Notation The numerical value of G was first determined by Henry Cavendish in 1798. In S.I. units, G is $6.67 \times 10^{-11} \, \text{kg}^{-1} \, \text{m}^3 \, \text{s}^{-2}$ Example 5 demonstrates the extremely small gravitational attraction between two everyday objects. You can ignore the gravitational force between small objects in your calculations.

By modelling the Earth as a sphere of mass M and radius R and using Newton's law of gravitation

$$F = \frac{GmM}{R^2} \quad \text{and} \quad F = mg$$

so $\quad \dfrac{GmM}{R^2} = mg$

and hence $\quad G = \dfrac{gR^2}{M}$

This relationship means you can answer questions involving gravity without using G explicitly.

When a particle is moving away from or towards the Earth, the distance d between the two particles is changing. As the force of attraction between them is given by $F = \dfrac{GM_1M_2}{d^2}$ it follows that the force is a function of displacement. Therefore, the methods of Section 3.1 must be used to solve problems.

Example **5**  **SKILLS** PROBLEM-SOLVING

Two particles of masses $0.5\,\text{kg}$ and $2.5\,\text{kg}$ are $4\,\text{cm}$ apart. Calculate the magnitude of the gravitational force between them.

$F = \dfrac{GM_1M_2}{d^2}$

$\quad = \dfrac{6.67 \times 10^{-11} \times 0.5 \times 2.5}{0.04^2}$

$\quad = 5.210\ldots \times 10^{-8}$

The magnitude of the gravitational force between the particles is $5.21 \times 10^{-8}\,\text{N}$ (3 s.f.).

Use $F = \dfrac{GM_1M_2}{d^2}$ with

$G = 6.67 \times 10^{-11}$, $M_1 = 0.5$, $M_2 = 2.5$ and $d = 0.04$

This value is so small that it can be ignored in most calculations.

Example **6** **SKILLS** CRITICAL THINKING

Above the Earth's surface, the magnitude of the force on a particle due to the Earth's gravitational force is inversely proportional to the square of the distance of the particle from the centre of the Earth. The acceleration due to gravity on the surface of the Earth is g and the Earth can be modelled as a sphere of radius R. A particle P of mass m is a distance $(x - R)$, (where $x \geqslant R$), above the surface of the Earth.

a Prove that the magnitude of the gravitational force acting on P is $\dfrac{mgR^2}{x^2}$

A spacecraft S is fired vertically upwards from the surface of the Earth. When it is at a height $2R$ above the surface of the Earth its speed is $\dfrac{1}{2}\sqrt{gR}$. Assuming that air resistance can be ignored and that the rocket's engine is turned off immediately after the rocket is fired,

b find, in terms of g and R, the speed with which S was fired.

a $F \propto \dfrac{1}{x^2}$ or $F = \dfrac{k}{x^2}$

So on the surface of the Earth $F = \dfrac{k}{R^2}$

On the surface of the Earth the magnitude of the force $= mg$ ———— The force is the weight of the particle.

$\therefore mg = \dfrac{k}{R^2} \Rightarrow k = mgR^2$

$\therefore$ the magnitude of the gravitational force is $\dfrac{mgR^2}{x^2}$

b $F = ma$

$\dfrac{mgR^2}{x^2} = -ma$ ———— Taking upwards as the positive direction, the force acts vertically downwards.

$v\dfrac{dv}{dx} = -\dfrac{gR^2}{x^2}$ ———— The force is a function of x so replace a with $v\dfrac{dv}{dx}$

$\int v\,dv = -gR^2\int \dfrac{1}{x^2}\,dx$

$\dfrac{1}{2}v^2 = gR^2\dfrac{1}{x} + C$ ———— Separate the variables and integrate. Don't forget to include the constant of integration.

$x = 3R, \; v = \dfrac{1}{2}\sqrt{gR}$

$\Rightarrow \dfrac{1}{2} \times \dfrac{1}{4}gR = gR^2 \times \dfrac{1}{3R} + C$

$C = -\dfrac{5}{24}gR$ ———— Use the information in the question to obtain the value of C.

$\dfrac{1}{2}v^2 = gR^2\dfrac{1}{x} - \dfrac{5}{24}gR$

$x = R \Rightarrow \dfrac{1}{2}v^2 = gR^2 \times \dfrac{1}{R} - \dfrac{5}{24}gR = \dfrac{19}{24}gR$ ———— Finally make $x = R$ and obtain an expression for v as required.

$\therefore S$ is fired with speed $\sqrt{\dfrac{19}{12}gR}$

Exercise **3B**   **SKILLS**   **CRITICAL THINKING**

(P) **1** Above the Earth's surface, the magnitude of the force on a particle due to the Earth's gravitational force is inversely proportional to the square of the distance of the particle from the centre of the Earth. The acceleration due to gravity on the surface of the Earth is g and the Earth can be modelled as a sphere of radius R. A particle P of mass m is a distance $(x - R)$ (where $x \geqslant R$) above the surface of the Earth. Prove that the magnitude of the gravitational force acting on P is $\dfrac{mgR^2}{x^2}$

(P) **2** The Earth can be modelled as a sphere of radius R. At a distance x (where $x \geqslant R$) from the centre of the Earth the magnitude of the acceleration due to the Earth's gravitational force is A. On the surface of the Earth, the magnitude of the acceleration due to the Earth's gravitational force is g. Prove that $A = \dfrac{gR^2}{x^2}$

(E/P) **3** A spacecraft S is fired vertically upwards from the surface of the Earth. When it is at a height R where R is the radius of the Earth, above the surface of the Earth its speed is $\sqrt{gR}$. Model the spacecraft as a particle and the Earth as a sphere of radius R and find, in terms of g and R, the speed with which S was fired. (You may assume that air resistance can be ignored and that the rocket's engine is turned off immediately after the rocket is fired.) **(7 marks)**

Watch out In questions 3 to 6 you may assume either of the results proved in questions 1 and 2.

(P) **4** A rocket of mass m is fired vertically upwards from the surface of the Earth with initial speed U. The Earth is modelled as a sphere of radius R and the rocket as a particle. Find an expression for the speed of the rocket when it has travelled a distance X metres. (You may assume that air resistance can be ignored and that the rocket's engine is turned off immediately after the rocket is fired.)

(P) **5** A particle is fired vertically upwards from the Earth's surface. The initial speed of the particle is u where $u^2 = 3gR$ and R is the radius of the Earth. Find, in terms of g and R, the speed of the particle when it is at a height $4R$ above the Earth's surface. (You may assume that air resistance can be ignored.)

(P) **6** A particle is moving in a straight line towards the centre of the Earth, which is assumed to be a sphere of radius R. The particle starts from rest when its distance from the centre of the Earth is $3R$. Find the speed of the particle as it hits the surface of the Earth. (You may assume that air resistance can be ignored.)

(E/P) **7** A space shuttle S of mass m moves in a straight line towards the centre of the Earth. The Earth is modelled as a sphere of radius R and S is modelled as a particle. When S is at a distance x $(x \geqslant R)$ from the centre of the Earth, the gravitational force **exerted** by the Earth on S is directed towards the centre of the Earth. The magnitude of this force is inversely proportional to x^2.

a Prove that the magnitude of the gravitational force on S is $\dfrac{mgR^2}{x^2}$

When S is at a height of $3R$ above the surface of the Earth, the speed of S is $\sqrt{2gR}$
Assuming that air resistance can be ignored,

b find, in terms of g and R, the speed of S as it hits the surface of the Earth. **(7 marks)**

Challenge

Given that $G = 6.67 \times 10^{-11}\ \text{kg}^{-1}\,\text{m}^3\,\text{s}^{-2}$, $g = 9.81\ \text{m s}^{-2}$ and that the radius of the Earth is 6.3781×10^6 m, estimate

a the mass of the Earth

b the average **density** of the Earth.

3.3 Simple harmonic motion

You can solve problems about a particle which is moving in a straight line with **simple harmonic motion**.

- **Simple harmonic motion (S.H.M.) is motion in which the acceleration of a particle P is always towards a fixed point O on the line of motion of P and has magnitude proportional to the displacement of P from O.**

> **Notation**   The point O is called the **centre of oscillation**.

> **Online**   Explore simple harmonic motion using GeoGebra.

We write $\ddot{x} = -\omega^2 x$ ———————————   The minus sign means that the acceleration is always directed towards O.

This can be shown on a diagram.

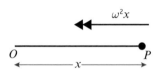

As $\ddot{x}$ is a function of x, we use $\ddot{x} = v\dfrac{\mathrm{d}v}{\mathrm{d}x}$ to derive an expression for the velocity of P.

$$v\frac{\mathrm{d}v}{\mathrm{d}x} = -\omega^2 x$$

$$\int v\,\mathrm{d}v = \int -\omega^2 x\,\mathrm{d}x$$

$$\frac{1}{2}v^2 = -\omega^2 \times \frac{1}{2}x^2 + C$$

The speed of P is the modulus of v or the modulus of $\dfrac{\mathrm{d}x}{\mathrm{d}t}$

This speed is zero when x has its maximum or minimum value.
Let the maximum displacement of P from O be a. This gives

$$0 = -\omega^2 \times \frac{1}{2}a^2 + C$$

$$C = \frac{1}{2}\omega^2 a^2$$

Hence $\dfrac{1}{2}v^2 = -\dfrac{1}{2}\omega^2 x^2 + \dfrac{1}{2}\omega^2 a^2$

or $v^2 = \omega^2(a^2 - x^2)$

You can derive an expression for the displacement of P at time t by considering $v = \omega\sqrt{a^2 - x^2}$

First consider that $\dfrac{\mathrm{d}x}{\mathrm{d}t} = \omega\sqrt{a^2 - x^2}$, then $\displaystyle\int \frac{\mathrm{d}x}{\sqrt{a^2 - x^2}} = \omega\int\mathrm{d}t$

Using the substitution $x = a\sin\theta$ with $\mathrm{d}x = a\cos\theta\,\mathrm{d}\theta$ leads to $\displaystyle\int \frac{a\cos\theta}{\sqrt{a^2 - a^2\sin^2\theta}}\,\mathrm{d}\theta = \omega\int\mathrm{d}t$

Then $\displaystyle\int\mathrm{d}\theta = \omega\int\mathrm{d}t$

Integrating gives $\theta = \omega t + \alpha$, then $\sin\theta = \sin(\omega t + \alpha)$

Since $\frac{x}{a} = \sin\theta$ this leads to $x = a\sin(\omega t + \alpha)$

So the motion of the particle is a sine function with maximum and minimum values $\pm a$ and **period** $\frac{2\pi}{\omega}$. The value a is called the **amplitude** of the motion, and $-a \leqslant x \leqslant a$ for all values of t.

We can use different values of α to investigate the motion further.

Case 1: $\alpha = 0$

When $\alpha = 0$, the equation $x = a\sin(\omega t + \alpha)$ becomes $x = a\sin\omega t$

Here is the graph of $x = a\sin\omega t$

As the graph passes through the origin, we have $x = 0$ when $t = 0$. Thus $x = a\sin\omega t$ gives the displacement from the centre of oscillation of a particle moving with S.H.M. of amplitude a and period $\frac{2\pi}{\omega}$ which is at the centre of the oscillation and moving in the positive direction when $t = 0$

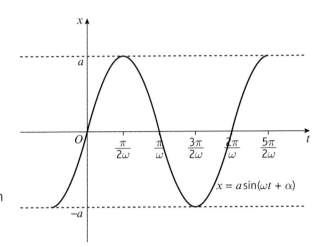

Case 2: $\alpha = \frac{\pi}{2}$

When $\alpha = \frac{\pi}{2}$, the equation $x = a\sin(\omega t + \alpha)$ becomes $x = a\sin\left(\omega t + \frac{\pi}{2}\right)$

The graph of $x = a\sin\left(\omega t + \frac{\pi}{2}\right)$ is

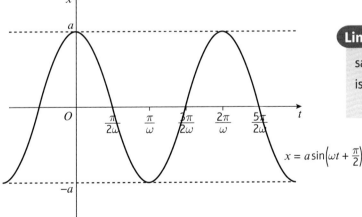

Links The graph of $x = a\sin\left(\omega t + \frac{\pi}{2}\right)$ is the same shape as the graph of $x = a\sin\omega t$ but is translated $\frac{\pi}{2\omega}$ to the left.

← Pure 1 Section 4.4

This is also the graph of $x = a\cos\omega t$

When $t = 0$ the particle's displacement from O is a.

Once again, the amplitude is a and the period is $\frac{2\pi}{\omega}$

Case 3: α is neither of the above.

When α takes some value other than 0 or $\frac{\pi}{2}$ the graph of $x = a \sin (\omega t + \alpha)$ is a **translation** of the graph of $x = a \sin \omega t$ through a distance $\frac{\alpha}{\omega}$ to the left.

The particle is neither at the centre nor at an extreme point of the oscillation when $t = 0$

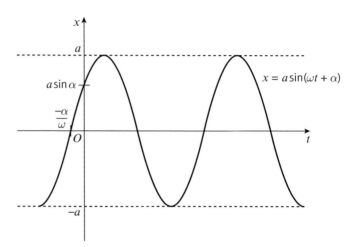

For S.H.M. of amplitude a defined by the equation $\ddot{x} = -\omega^2 x$

- $v^2 = \omega^2(a^2 - x^2)$
- If P is at the centre of the oscillation when $t = 0$, use $x = a \sin \omega t$
- If P is at an end point of the oscillation when $t = 0$, use $x = a \cos \omega t$
- If P is at some other point when $t = 0$, use $x = a \sin (\omega t + \alpha)$
- The period of the oscillation is $T = \dfrac{2\pi}{\omega}$

Notation These are standard results which may be quoted, without proof, in your exam.

Geometrical methods

You can also use a geometrical method to solve simple harmonic motion problems.

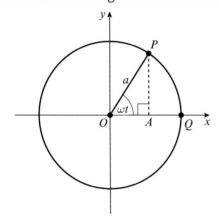

A particle P is moving round a circle of radius a, centre the origin O. The particle has a constant **angular** speed ω in an anticlockwise sense. The foot of the perpendicular from P to the x-axis is the point A.

Links Angular speed, ω, is the rate at which the radius is turning, measured in rad s^{-1}. → **Mechanics 3 Section 4.1**

The motion is timed from the instant when P is at the point Q on the x-axis. So t seconds later, $\angle POA = \omega t$ and $OA = x = a \cos \omega t$

Hence $\dot{x} = -a\omega \sin \omega t$

and $\ddot{x} = -a\omega^2 \cos \omega t = -\omega^2 x$

This shows that the point A is moving along the x-axis with simple harmonic motion.

The amplitude is a and the period is $\frac{2\pi}{\omega}$

The circle associated with any particular simple harmonic motion is called the **reference circle**. Using the reference circle can be useful when calculating the time taken for a particle to move between two points of the oscillation, as shown in Example 11.

Example **7**  **SKILLS** PROBLEM-SOLVING

A particle is moving along a straight line with S.H.M. The amplitude of the motion is $0.8\,\text{m}$. It passes through the centre of the oscillation O with speed $2\,\text{m}\,\text{s}^{-1}$. Calculate

a the period of the oscillation

b the speed of the particle when it is $0.4\,\text{m}$ from O

c the time the particle takes to travel $0.4\,\text{m}$ from O.

a $v^2 = \omega^2(a^2 - x^2)$ ——— Use $v^2 = \omega^2(a^2 - x^2)$ with $v = 2$, $a = 0.8$ and $x = 0$

$2^2 = \omega^2(0.8^2 - 0)$

$\omega = \dfrac{2}{0.8} = 2.5$ ——— Solve for ω.

$\text{period} = \dfrac{2\pi}{\omega}$

$\qquad = \dfrac{2\pi}{2.5} = \dfrac{4\pi}{5}$ ——— Use $\text{period} = \dfrac{2\pi}{\omega}$ with $\omega = 2.5$

The period is $\dfrac{4\pi}{5}\,\text{s}$.

b $v^2 = \omega^2(a^2 - x^2)$

$v^2 = 2.5^2(0.8^2 - 0.4^2)$ ——— Use $v^2 = \omega^2(a^2 - x^2)$ with $\omega = 2.5$ $a = 0.8$ and $x = 0.4$ and solve for v.

$v = 1.732\ldots$

The particle's speed is $1.73\,\text{m}\,\text{s}^{-1}$ (3 s.f.).

c $x = a\sin\omega t$

$0.4 = 0.8\sin 2.5t$ ——— Take $t = 0$ at the centre and use $x = a\sin\omega t$ with $\omega = 2.5$ $a = 0.8$ and $x = 0.4$

$\sin 2.5t = 0.5$

$2.5t = \arcsin 0.5$

$t = \dfrac{1}{2.5}\arcsin 0.5 = 0.2094\ldots$

Watch out Remember to have your calculator in radian mode.

The particle takes $0.209\,\text{s}$ (3 s.f.).

Example **8** **SKILLS** PROBLEM-SOLVING

A particle P of mass $0.5\,\text{kg}$ is moving along a straight line. At time t seconds, the distance of P from a fixed point O on the line is x metres. The force acting on P has magnitude $10x$ and acts in the direction PO.

a Show that P is moving with simple harmonic motion.

b Find the period of the motion.

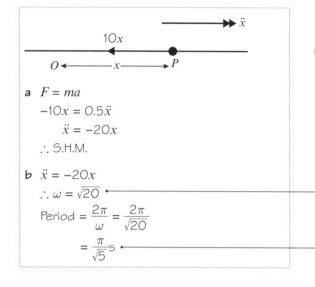

a $F = ma$

$-10x = 0.5\ddot{x}$

$\ddot{x} = -20x$

$\therefore$ S.H.M.

b $\ddot{x} = -20x$

$\therefore \omega = \sqrt{20}$

Period $= \dfrac{2\pi}{\omega} = \dfrac{2\pi}{\sqrt{20}}$

$\quad\quad\quad = \dfrac{\pi}{\sqrt{5}}\,s$

Problem-solving

If you need to show that a particle is moving with simple harmonic motion you must show that $\ddot{x} = -\omega^2 x$ and state the conclusion.

Compare the equation for P found in **a** with $\ddot{x} = -\omega^2 x$ to find ω.

The answer can be left in its exact form or given correct to 3 s.f. (1.40 s).

Example **9**   **SKILLS**   PROBLEM-SOLVING

A particle is moving in a straight line with simple harmonic motion. Its maximum acceleration is $12\,\mathrm{m\,s^{-2}}$ and its maximum speed is $4\,\mathrm{m\,s^{-1}}$. Calculate

a the period of the motion

b the amplitude of the motion

c the length of time, during one complete oscillation, that the particle is within $\dfrac{1}{3}$ m of the centre of the oscillation.

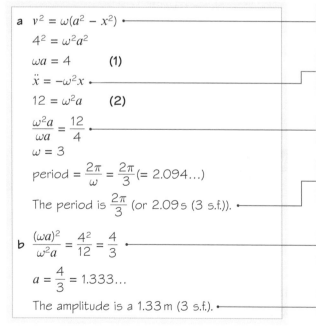

a $v^2 = \omega(a^2 - x^2)$

$4^2 = \omega^2 a^2$

$\omega a = 4$ (1)

$\ddot{x} = -\omega^2 x$

$12 = \omega^2 a$ (2)

$\dfrac{\omega^2 a}{\omega a} = \dfrac{12}{4}$

$\omega = 3$

period $= \dfrac{2\pi}{\omega} = \dfrac{2\pi}{3}(= 2.094...)$

The period is $\dfrac{2\pi}{3}\,s$ (or 2.09 s (3 s.f.)).

b $\dfrac{(\omega a)^2}{\omega^2 a} = \dfrac{4^2}{12} = \dfrac{4}{3}$

$a = \dfrac{4}{3} = 1.333...$

The amplitude is a 1.33 m (3 s.f.).

As $v^2 = \omega^2(a^2 - x^2)$, v is maximum when $x = 0$

As $\ddot{x} = -\omega^2 x$, and hence $|\ddot{x}| = |-\omega^2 x| = \omega^2|x|$ acceleration is maximum when $|x| = a$

Divide equation **(2)** by equation **(1)** to find ω.

Unless you are told the accuracy required, give the exact answer.

Square equation **(1)** and divide by equation **(2)** to find a.

The exact answer, $\dfrac{4}{3}$ or $1\dfrac{1}{3}$, is acceptable.

c Taking the initial displacement from O as 0.

$x = a \sin \omega t$ so $x = \frac{4}{3} \sin 3t$

Need to find t such that $-\frac{1}{3} \leqslant \frac{4}{3} \sin 3t \leqslant \frac{1}{3}$

$\frac{4}{3} \sin 3t = \frac{1}{3}$

$\sin 3t = \frac{1}{4} \Rightarrow t = 0.08423...$

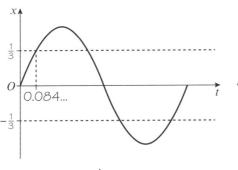

Total time within $\frac{1}{3}$ m $= 4 \times 0.08423...$

$\qquad\qquad\qquad = 0.337$ s (3 s.f.).

> **Problem-solving**
>
> The initial displacement does not affect the answer, so take $x = 0$ when $t = 0$ to simplify the expression for the displacement of the particle.

Choose the first positive value of t that satisfies the equation, and then draw a sketch to determine the total time that the particle is within $\frac{1}{3}$ m of O. You could work out all the solutions of $\frac{4}{3} \sin 3t = \pm\frac{1}{3}$ in one complete oscillation, but it is easier to use **symmetry**.

Example 10 **SKILLS** INTERPRETATION

A small rowing boat is floating on the surface of the sea, tied to a pier. The boat moves up and down in a vertical line and it can be modelled as a particle moving with simple harmonic motion. The boat takes 2 s to travel directly from its highest point, which is 3 m below the pier, to its lowest point. The maximum speed of the boat is 3 m s⁻¹. Calculate

a the amplitude of the motion

b the time taken by the boat to rise from its lowest point to a point 5 m below the pier.

a Period $= \frac{2\pi}{\omega} = 4$

$\Rightarrow \omega = \frac{2\pi}{4} = \frac{\pi}{2}$

$v^2 = \omega^2(a^2 - x^2)$

$3^2 = \left(\frac{\pi}{2}\right)^2 (a^2 - 0)$

$a^2 = 3^2 \times \left(\frac{2}{\pi}\right)^2$

$a = \frac{6}{\pi} = 1.909...$

The amplitude is 1.91 m (3 s.f.).

The time taken from the highest point to the lowest point is half the period.

The maximum speed of the boat occurs when it passes the centre of oscillation, or when $x = 0$

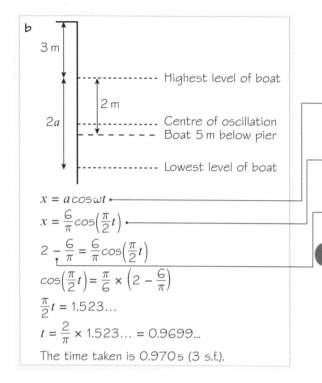

Time from the lowest point is needed so use
$x = a\cos\omega t$

Using the exact values of a and ω gives a more accurate answer.

$x = a\cos\omega t$

$x = \dfrac{6}{\pi}\cos\left(\dfrac{\pi}{2}t\right)$

$2 - \dfrac{6}{\pi} = \dfrac{6}{\pi}\cos\left(\dfrac{\pi}{2}t\right)$

$\cos\left(\dfrac{\pi}{2}t\right) = \dfrac{\pi}{6} \times \left(2 - \dfrac{6}{\pi}\right)$

$\dfrac{\pi}{2}t = 1.523\ldots$

$t = \dfrac{2}{\pi} \times 1.523\ldots = 0.9699\ldots$

The time taken is 0.970 s (3 s.f.).

x is measured from the centre of the oscillation. The diagram shows that $x = (2 - a)$ m.

Problem-solving

This analysis considers down as the positive direction, to simplify the working. If you want to consider up as the positive direction you could write the displacement as

$x = -\dfrac{6}{\pi}\cos\left(\dfrac{\pi}{2}t\right)$ and set $x = \dfrac{6}{\pi} - 2$

Example **11**   **SKILLS**   **CRITICAL THINKING**

A particle P is moving with simple harmonic motion along a straight line. The centre of the oscillation is the point O, the amplitude is 0.6 m and the period is 8 s. The points A and B are points on the line of P's motion and are on opposite sides of O. The distance OA is 0.5 m and OB is 0.1 m. Calculate the time taken by P to move directly from A to B.

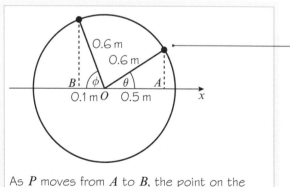

Draw a diagram showing a reference circle of radius 0.6 m. The diameter of this circle represents the line on which P is moving. Show points A and B on this diameter.

Online   Explore calculations for simple harmonic motion with a reference circle using GeoGebra.

As P moves from A to B, the point on the circle moves round an arc of the circle which subtends an angle of $\pi - \theta - \phi$ at O.

$\theta = \arccos\left(\dfrac{0.5}{0.6}\right) = \arccos\left(\dfrac{5}{6}\right)$

$\phi = \arccos\left(\dfrac{0.1}{0.6}\right) = \arccos\left(\dfrac{1}{6}\right)$

Use the diagram to find θ and ϕ. Keep the exact forms.

Time $= \dfrac{\pi - \theta - \phi}{2\pi} \times 8$

The period is 8 s. This corresponds to an angle of 2π at the centre. Use proportion to find the time taken.

$= \dfrac{\pi - \arccos\left(\dfrac{5}{6}\right) - \arccos\left(\dfrac{1}{6}\right)}{2\pi} \times 8$

$= 1.467\ldots$

The time to travel from A to B is 1.47 s (3 s.f.).

Exercise 3C  SKILLS PROBLEM-SOLVING

1 A particle P is moving in a straight line with simple harmonic motion. The amplitude of the oscillation is 0.5 m and P passes through the centre of the oscillation O with speed 2 m s^{-1}. Calculate
   a the period of the oscillation
   b the speed of P when $OP = 0.2$ m.

2 A particle P is moving in a straight line with simple harmonic motion. The period is $\dfrac{\pi}{3}$ s and P's maximum speed is 6 m s^{-1}. The centre of the oscillation is O. Calculate
   a the amplitude of the motion
   b the speed of P 0.3 s after passing through O.

3 A particle is moving in a straight line with simple harmonic motion. Its maximum speed is 10 m s^{-1} and its maximum acceleration is 20 m s^{-2}. Calculate
   a the amplitude of the motion
   b the period of the motion.

4 A particle is moving in a straight line with simple harmonic motion. The period of the motion is $\dfrac{3\pi}{5}$ s and the amplitude is 0.4 m. Calculate the maximum speed of the particle.

5 A particle is moving in a straight line with simple harmonic motion. Its maximum acceleration is 15 m s^{-2} and its maximum speed is 18 m s^{-1}. Calculate the speed of the particle when it is 2.5 m from the centre of the oscillation.

6 A particle P is moving in a straight line with simple harmonic motion. The centre of the oscillation is O and the period is $\dfrac{\pi}{2}$ s. When $OP = 1.2$ m, P has speed 1.5 m s^{-1}.
   a Find the amplitude of the motion.
   At time t seconds the displacement of P from O is x metres. When $t = 0$, P is passing through O.
   b Find an expression for x in terms of t.

7 A particle is moving in a straight line with simple harmonic motion. The particle performs 6 complete oscillations per second and passes through the centre of the oscillation, O, with speed $5\,\mathrm{m\,s^{-1}}$. When P passes through the point A the magnitude of P's acceleration is $20\,\mathrm{m\,s^{-1}}$. Calculate

 a the amplitude of the motion

 b the distance OA.

(P) 8 A particle P is moving on a straight line with simple harmonic motion between two points A and B. The midpoint of AB is O. When $OP = 0.6\,\mathrm{m}$, the speed of P is $3\,\mathrm{m\,s^{-1}}$ and when $OP = 0.2\,\mathrm{m}$ the speed of P is $6\,\mathrm{m\,s^{-1}}$. Find

 a the distance AB

 b the period of the motion.

9 A particle is moving in a straight line with simple harmonic motion. When the particle is $1\,\mathrm{m}$ from the centre of the oscillation, O, its speed is $0.1\,\mathrm{m\,s^{-1}}$. The period of the motion is 2π seconds. Calculate

 a the maximum speed of the particle

 b the speed of the particle when it is $0.4\,\mathrm{m}$ from O.

(P) 10 A piston of mass $1.2\,\mathrm{kg}$ is moving with simple harmonic motion inside a cylinder. The distance between the end points of the motion is $2.5\,\mathrm{m}$ and the piston is performing 30 complete oscillations per minute. Calculate the maximum value of the kinetic energy of the piston.

11 A marker buoy is moving in a vertical line with simple harmonic motion. The buoy rises and falls through a distance of $0.8\,\mathrm{m}$ and takes $2\,\mathrm{s}$ for each complete oscillation. Calculate

 a the maximum speed of the buoy

 b the time taken for the buoy to fall a distance $0.6\,\mathrm{m}$ from its highest point.

(E) 12 Points O, A and B lie in that order in a straight line. A particle P is moving on the line with simple harmonic motion. The motion has period $2\,\mathrm{s}$ and amplitude $0.5\,\mathrm{m}$. The point O is the centre of the oscillation, $OA = 0.2\,\mathrm{m}$ and $OB = 0.3\,\mathrm{m}$. Calculate the time taken by P to move directly from A to B. **(5 marks)**

(E/P) 13 A particle P is moving along the x-axis. At time t seconds the displacement, x metres, of P from the origin O is given by $x = 4\sin 2t$

 a Prove that P is moving with simple harmonic motion. **(5 marks)**

 b Write down the amplitude and period of the motion. **(2 marks)**

 c Calculate the maximum speed of P. **(3 marks)**

 d Calculate the least value of t $(t > 0)$ for which P's speed is $4\,\mathrm{m\,s^{-1}}$. **(3 marks)**

 e Calculate the least value of t $(t > 0)$ for which $x = 2$ **(2 marks)**

(E/P) **14** A particle P is moving along the x-axis. At time t seconds the displacement, x metres, of P

from the origin O is given by $x = 3\sin\left(4t + \dfrac{1}{2}\right)$

    a Prove that P is moving with simple harmonic motion.   **(5 marks)**

    b Write down the amplitude and period of the motion.   **(2 marks)**

    c Calculate the value of x when $t = 0$ **(2 marks)**

    d Calculate the value of t $(t > 0)$ the first time P passes through O.   **(3 marks)**

(E/P) **15** On a certain day, low tide in a harbour is at 10 a.m. and the depth of the water is 5 m. High tide on the same day is at 4.15 p.m. and the water is then 15 m deep. A ship which needs a depth of water of 7 m needs to enter the harbour. Assuming that the water can be modelled as rising and falling with simple harmonic motion, calculate

    a the earliest time, to the nearest minute, after 10 a.m. at which the ship can enter the
harbour   **(4 marks)**

    b the time by which the ship must leave.   **(3 marks)**

(E/P) **16** Points A, O and B lie in that order in a straight line. A particle P is moving on the line with simple harmonic motion with centre O. The period of the motion is 4 s and the amplitude is 0.75 m. The distance OA is 0.4 m and AB is 0.9 m. Calculate the time taken by P to move directly from B to A.   **(5 marks)**

Challenge

A particle P is moving along the x-axis with simple harmonic motion. The origin O is the centre of oscillation. When the displacements from O are x_1 and x_2 the particle has speeds of v_1 and v_2 respectively (i.e. in the same order as things already mentioned). Find the period of the motion in terms of x_1 v_1 and v_2

3.4 Horizontal oscillation

You can investigate the motion of a particle which is attached to an elastic spring or string and is oscillating in a horizontal line.

If an elastic spring has one end attached to a fixed point A on a smooth horizontal surface a particle P can be attached to the free end. When P is pulled away from A and released P will move towards A.

Hooke's law: $T = \dfrac{\lambda x}{l}$

         $F = ma$

         $-T = m\ddot{x}$

         $m\ddot{x} = -\dfrac{\lambda x}{l}$

         $\ddot{x} = -\dfrac{\lambda}{ml}x$

Links λ is the modulus of elasticity of the spring and l is its natural length.

← **Mechanics 3 Section 2.1**

λ, m and l are all positive constants, so the equation is of the form $\ddot{x} = -\omega^2 x$

So P is moving with S.H.M.

The initial extension is the maximum value of x, so is the same as the amplitude.

When the particle is attached to an elastic **spring**, the particle will perform complete oscillations. This is because there will always be a force acting – a tension when the spring is stretched and a thrust when the spring is compressed. The centre of the oscillation is where the tension is zero; that is the point when the spring has returned to its natural length.

When the particle is attached to an elastic **string**, the particle will move with S.H.M. only while the string is taut. Once the string becomes slack there is no tension and the particle continues to move with constant speed until the string becomes taut again.

For a particle moving on a smooth horizontal surface attached to one end of an elastic spring
- the particle will move with S.H.M.
- the particle will perform complete oscillations.

For a particle moving on a smooth horizontal surface attached to one end of an elastic string
- the particle will move with S.H.M. while the string is taut
- the particle will move with constant speed while the string is slack.

To solve problems involving elastic springs and strings
- use Hooke's law to find the tension
- use $F = ma$ to obtain ω
- use information given in the question to obtain the amplitude.

Sometimes the particle is attached to two springs or strings which are stretched between two fixed points. When this happens you will need to find the tensions in both the springs or strings.

Example **12**   **SKILLS** ▷ INTERPRETATION

A particle P of mass $0.6\,\text{kg}$ rests on a smooth horizontal floor attached to one end of a light elastic string of natural length $0.8\,\text{m}$ and modulus of elasticity $16\,\text{N}$. The other end of the string is fixed to a point A on the floor. The particle is pulled away from A until AP measures $1.2\,\text{m}$ and released.

a Show that, while the string remains taut, P moves with simple harmonic motion.

b Calculate the speed of P when the string returns to its natural length.

c Calculate the time that elapses between the point where the string becomes slack and the point where it next becomes taut.

d Calculate the time taken by the particle to return to its starting point for the first time.

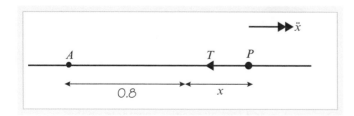

a $T = \dfrac{\lambda x}{l}$

$T = \dfrac{16x}{0.8} = 20x$

$F = ma$

$20x = -0.6\ddot{x}$

$\ddot{x} = -\dfrac{20}{0.6}x = -\dfrac{100}{3}x$

The particle is moving with S.H.M.

Use Hooke's law with $\lambda = 16$ and $l = 0.8$ to find the tension.

Use $F = ma$ with $F = T = 20x$ and $m = 0.6$ Remember that the positive direction is the direction of x increasing, and that the acceleration acts in the opposite direction.

The equation reduces to the form $\ddot{x} = -\omega^2 x$, so S.H.M. is proved.

b $\ddot{x} = -\dfrac{100}{3}x$

$\omega^2 = \dfrac{100}{3}$

$v^2 = \omega^2(a^2 - x^2)$

$v^2 = \dfrac{100}{3} \times 0.4^2$

$v = 2.309...$

At the natural length P has speed $2.31\,\text{ms}^{-1}$ (3 s.f.).

Compare the equation obtained in **a** with $\ddot{x} = -\omega^2 x$ to find ω^2.

The amplitude is the same as the initial extension and at the natural length $x = 0$

c The particle now moves a distance $1.6\,\text{m}$ at $2.309...\,\text{ms}^{-1}$.

Time taken $= \dfrac{1.6}{2.309...} = 0.6928...$

The string is slack for $0.693\,\text{s}$ (3 s.f.).

The particle moves at a constant speed while the string is slack.

d Period of the S.H.M. $= \dfrac{2\pi}{\omega} = 2\pi \times \sqrt{\dfrac{3}{100}} = 1.088...$

Total time $= 1.088... + (2 \times 0.6928...) = 2.473...$

The time taken is $2.47\,\text{s}$ (3 s.f.).

The particle moves through a complete oscillation and two intervals of constant speed (when the string is slack).

Example 13 **SKILLS** INTERPRETATION

A particle P of mass $0.8\,\text{kg}$ is attached to the ends of two identical light elastic springs of natural length $1.6\,\text{m}$ and modulus of elasticity $16\,\text{N}$. The free ends of the springs are attached to two points A and B which are $4\,\text{m}$ apart on a smooth horizontal surface.

The point C lies between A and B such that ABC is a straight line and $AC = 2.4\,\text{m}$

The particle is held at C and then released from rest.

a Show that the subsequent motion is simple harmonic motion.

b Find the period and amplitude of the motion.

c Calculate the maximum speed of P.

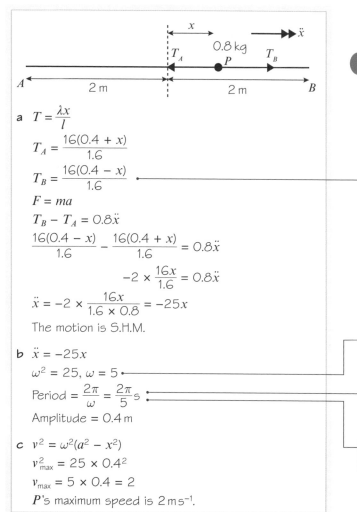

a $T = \dfrac{\lambda x}{l}$

$T_A = \dfrac{16(0.4 + x)}{1.6}$

$T_B = \dfrac{16(0.4 - x)}{1.6}$

$F = ma$

$T_B - T_A = 0.8\ddot{x}$

$\dfrac{16(0.4 - x)}{1.6} - \dfrac{16(0.4 + x)}{1.6} = 0.8\ddot{x}$

$\qquad\qquad -2 \times \dfrac{16x}{1.6} = 0.8\ddot{x}$

$\ddot{x} = -2 \times \dfrac{16x}{1.6 \times 0.8} = -25x$

The motion is S.H.M.

b $\ddot{x} = -25x$

$\omega^2 = 25,\ \omega = 5$

$\text{Period} = \dfrac{2\pi}{\omega} = \dfrac{2\pi}{5}\,\text{s}$

$\text{Amplitude} = 0.4\,\text{m}$

c $v^2 = \omega^2(a^2 - x^2)$

$v_{max}^2 = 25 \times 0.4^2$

$v_{max} = 5 \times 0.4 = 2$

P's maximum speed is $2\,\text{m s}^{-1}$.

Problem-solving

Use Hooke's law to find the tensions in each spring. Use your diagram to work out the extensions.

Use $F = ma$ to form an equation of motion for P. Reduce this to the form $\ddot{x} = -\omega^2x$ to establish S.H.M.

Compare the equation found in **a** with $\ddot{x} = -\omega^2x$ to find ω.

You can give an exact value or a 3 s.f. answer (1.26 s) for the period.

As the springs are identical the centre of the oscillation is at the midpoint of AB.

Exercise **3D**   **SKILLS**   INTERPRETATION

1 A particle P of mass $0.5\,\text{kg}$ is attached to one end of a light elastic spring of natural length $0.6\,\text{m}$ and modulus of elasticity $60\,\text{N}$. The other end of the spring is fixed to a point A on the smooth horizontal surface on which P rests. The particle is held at rest with $AP = 0.9\,\text{m}$ and then released.

  a Show that P moves with simple harmonic motion.

  b Find the period and amplitude of the motion.

  c Calculate the maximum speed of P.

2 A particle P of mass $0.8\,\text{kg}$ is attached to one end of a light elastic string of natural length $1.6\,\text{m}$ and modulus of elasticity $20\,\text{N}$. The other end of the string is fixed to a point O on the smooth horizontal surface on which P rests. The particle is held at rest with $OP = 2.6\,\text{m}$ and then released.

  a Show that, while the string is taut, P moves with simple harmonic motion.

  b Calculate the time from the instant of release until P returns to its starting point for the first time.

3 A particle P of mass $0.4\,\text{kg}$ is attached to one end of a light elastic string of modulus of elasticity $24\,\text{N}$ and natural length $1.2\,\text{m}$. The other end of the string is fixed to a point A on the smooth horizontal table on which P rests. Initially P is at rest with $AP = 1\,\text{m}$

The particle receives an impulse of magnitude $1.8\,\text{N s}$ in the direction AP.

a Show that, while the string is taut, P moves with simple harmonic motion.

b Calculate the time that elapses between the moment P receives the impulse and the next time the string becomes slack.

The particle comes to instantaneous rest for the first time at the point B.

c Calculate the distance AB.

(P) 4 A particle P of mass $0.8\,\text{kg}$ is attached to one end of a light elastic spring of natural length $1.2\,\text{m}$ and modulus of elasticity $80\,\text{N}$. The other end of the spring is fixed to a point O on the smooth horizontal surface on which P rests. The particle is held at rest with $OP = 0.6\,\text{m}$ and then released.

a Show that P moves with simple harmonic motion.

b Find the period and amplitude of the motion.

c Calculate the maximum speed of P.

5 A particle P of mass $0.6\,\text{kg}$ is attached to one end of a light elastic spring of modulus of elasticity $72\,\text{N}$ and natural length $1.2\,\text{m}$. The other end of the spring is fixed to a point A on the smooth horizontal table on which P rests. Initially P is at rest with $AP = 1.2\,\text{m}$. The particle receives an impulse of magnitude $3\,\text{N s}$ in the direction AP. Given that t seconds after the impulse the displacement of P from its initial position is x metres

a find an equation for x in terms of t

b calculate the maximum magnitude of the acceleration of P.

(P) 6 A particle of mass $0.9\,\text{kg}$ rests on a smooth horizontal surface attached to one end of a light elastic string of natural length $1.5\,\text{m}$ and modulus of elasticity $24\,\text{N}$. The other end of the string is attached to a point on the surface. The particle is pulled so that the string measures $2\,\text{m}$ and released from rest.

a State the amplitude of the resulting oscillation.

b Calculate the speed of the particle when the string becomes slack.

Before the string becomes taut again the particle hits a vertical surface which is at right angles to the particle's direction of motion. The coefficient of restitution between the particle and the vertical surface is $\dfrac{3}{5}$

c Calculate

  i the period

  ii the amplitude

of the oscillation which takes place when the string becomes taut once more.

(P) 7 A smooth cylinder is fixed with its axis horizontal. A piston of mass $2.5\,\text{kg}$ is inside the cylinder, attached to one end of the cylinder by a spring of modulus of elasticity $400\,\text{N}$ and natural length $50\,\text{cm}$. The piston is held at rest in the cylinder with the spring compressed to a length of $42\,\text{cm}$. The piston is then released. The spring can be modelled as a light elastic spring and the piston can be modelled as a particle.

a Find the period of the resulting oscillations.

b Find the maximum value of the kinetic energy of the piston.

(P) **8** A particle P of mass 0.5 kg is attached to one end of a light elastic string of natural length 0.4 m and modulus of elasticity 30 N. The other end of the string is attached to a point on the smooth horizontal surface on which P rests. The particle is pulled until the string measures 0.6 m and then released from rest.

    a Calculate the speed of P when the string becomes slack for the first time.

    The string breaks at the instant when it returns to its natural length for the first time. When P has travelled a distance 0.3 m from the point at which the string breaks the surface becomes rough. The coefficient of friction between P and the surface is 0.25. The particle comes to rest T seconds after it was released.

    b Find the value of T.

(E/P) **9** A particle P of mass 0.4 kg is attached to two identical light elastic springs each of natural length 1.2 m and modulus of elasticity 12 N. The free ends of the strings are attached to points A and B which are 4 m apart on a smooth horizontal surface. The point C lies between A and B with $AC = 1.4$ m and $CB = 2.6$ m. The particle is held at C and released from rest.

    a Show that P moves with simple harmonic motion. **(5 marks)**

    b Calculate the maximum value of the kinetic energy of P. **(3 marks)**

(E/P) **10** A particle P of mass m is attached to two identical light strings of natural length l and modulus of elasticity $3mg$. The free ends of the strings are attached to fixed points A and B which are $5l$ apart on a smooth horizontal surface. The particle is held at the point C, where $AC = l$ and A, B and C lie on a straight line, and is then released from rest.

    a Show that P moves with simple harmonic motion. **(3 marks)**

    b Find the period of the motion. **(2 marks)**

    c Write down the amplitude of the motion. **(4 marks)**

    d Find the speed of P when $AP = 3l$ **(2 marks)**

(E/P) **11** A light elastic string has natural length 2.5 m and modulus of elasticity 15 N. A particle P of mass 0.5 kg is attached to the string at the point K where K divides the unstretched string in the ratio 2 : 3. The ends of the string are then attached to the points A and B which are 5 m apart on a smooth horizontal surface. The particle is then pulled along AB and held at rest in contact with the surface at the point C where $AC = 3$ m and ACB is a straight line. The particle is then released from rest.

    a Show that P moves with simple harmonic motion of period $\frac{\pi}{5}\sqrt{2}$ **(5 marks)**

    b Find the amplitude of the motion. **(4 marks)**

3.5   Vertical oscillation

You can investigate the motion of a particle which is attached to an elastic spring or string and is oscillating in a vertical line.

A particle which is hanging in equilibrium attached to one end of an elastic spring or string, the other end of which is fixed, can be pulled downwards and released. The particle will then oscillate in a vertical line about its equilibrium position.

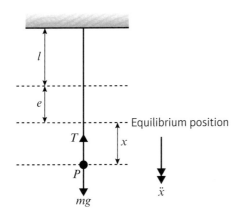

Taking downwards as the positive direction, when the particle is a distance x below its equilibrium position it has acceleration $\ddot{x}$.

At the equilibrium position, the tension in the spring or string is mg.

Using Hooke's law

$$T = \frac{\lambda \times \text{extension}}{l}$$

$$mg = \frac{\lambda e}{l}$$

$$e = \frac{mgl}{\lambda}$$

λ is the modulus of elasticity and l is the natural length of the spring or string. e is the extension of the spring or string in the equilibrium position.

Now consider the particle at a distance x below its equilibrium position.

$$T = \frac{\lambda \times \text{extension}}{l}$$

$$T = \frac{\lambda(x + e)}{l}$$

$$T = \frac{\lambda\left(x + \frac{mgl}{\lambda}\right)}{l} = \frac{\lambda x}{l} + mg$$

The particle is a distance x below the equilibrium position, so the extension is $x + e$

Using $F = ma$

$$mg - T = m\ddot{x}$$

$$mg - \left(\frac{\lambda x}{l} + mg\right) = m\ddot{x}$$

$$-\frac{\lambda x}{l} = m\ddot{x}$$

$$\ddot{x} = -\frac{\lambda}{ml}x$$

Watch out When using $F = ma$ the weight of the particle must be included as well as the tension.

λ, m and l are all positive constants, so the equation is of the form $\ddot{x} = -\omega^2 x$. It is the same result as obtained for a horizontal oscillation.

The particle is moving with S.H.M.

As in the case of horizontal oscillations, a particle attached to one end of a spring will perform complete oscillations. If the particle is attached to one end of an elastic string it will only move with S.H.M. while the string is taut. If the amplitude is greater than the extension at the equilibrium position, the string will become slack before the particle reaches the upper end of the oscillation. Once the string becomes slack, the oscillatory motion ceases and the particle moves freely under gravity until it falls back to the position where the string is once again taut.

- For a particle hanging in equilibrium attached to one end of an elastic spring and displaced vertically from its equilibrium position
  - the particle will move with simple harmonic motion.
  - the particle will perform complete oscillations
  - the centre of the oscillation will be the equilibrium position.
- For a particle hanging in equilibrium attached to one end of an elastic string and displaced vertically from its equilibrium position
  - the particle will move with simple harmonic motion while the string is taut
  - the particle will perform complete oscillations if the amplitude is no greater than the equilibrium extension
  - if the amplitude is greater than the equilibrium extension, the particle will move freely under gravity while the string is slack.

A particle can be attached to two springs or strings which are hanging side by side or stretched in a vertical line between two fixed points. The basic method of solution remains the same.

Example ⑭ **SKILLS** PROBLEM-SOLVING

A particle P of mass 1.2 kg is attached to one end of a light elastic spring of modulus of elasticity 60 N and natural length 60 cm. The other end of the spring is attached to a fixed point A on a ceiling. The particle hangs in equilibrium at the point B.

a Find the extension of the spring.

The particle is now raised vertically a distance of 15 cm and released from rest.

b Prove that P will move with simple harmonic motion.

c Find the period and amplitude of the motion.

d Find the speed of P as it passes through B.

e Find the speed of P at the instant when the spring has returned to its natural length.

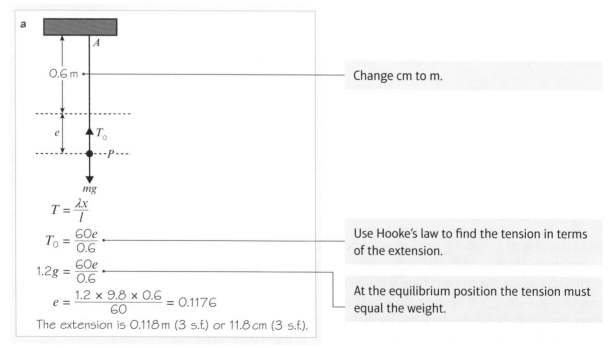

Change cm to m.

$$T = \frac{\lambda x}{l}$$

$$T_0 = \frac{60e}{0.6}$$

Use Hooke's law to find the tension in terms of the extension.

$$1.2g = \frac{60e}{0.6}$$

At the equilibrium position the tension must equal the weight.

$$e = \frac{1.2 \times 9.8 \times 0.6}{60} = 0.1176$$

The extension is 0.118 m (3 s.f.) or 11.8 cm (3 s.f.).

b

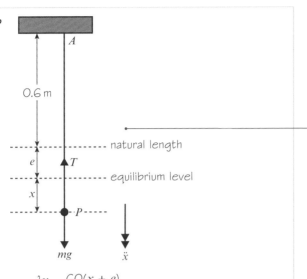

0.6 m

natural length

e T

equilibrium level

x

P

mg $\ddot{x}$

$T = \dfrac{\lambda x}{l} = \dfrac{60(x + e)}{1.2}$

$F = ma$

$mg - T = m\ddot{x}$

$mg - \dfrac{60(x + e)}{0.6} = m\ddot{x}$

$\dfrac{60e}{0.6} - \dfrac{60(x + e)}{0.6} = 1.2\ddot{x}$

$\ddot{x} = -\dfrac{60x}{1.2 \times 0.6} = -\dfrac{250}{3}x$

P moves with S.H.M.

c $\ddot{x} = -\dfrac{250}{3}x$

$\omega^2 = \dfrac{250}{3}$

$\text{period} = \dfrac{2\pi}{\omega} = \dfrac{2\pi}{\sqrt{\dfrac{250}{3}}} = 0.6882\ldots$

The period is 0.688 s (3 s.f.).
The amplitude is 15 cm.

d $v^2 = \omega^2(a^2 - x^2)$

$v_B^2 = \dfrac{250}{3}(0.15^2 - 0)$

$v_B = \sqrt{\dfrac{250}{3}} \times 0.15 = 1.369\ldots$

The speed at B is 1.37 m s⁻¹ (3 s.f.).

e $v^2 = \omega^2(a^2 - x^2)$

$v^2 = \dfrac{250}{3}(0.15^2 - (-0.1176)^2)$

$v = 0.8500\ldots$

The speed at the natural length is
0.850 m s⁻¹ (3 s.f.).

Online Explore the simple harmonic motion of a vertical spring using GeoGebra.

Draw a new diagram showing P at a distance x below the equilibrium level.

Use Hooke's law once more. This time the extension is $x + e$

Watch out When you use $F = ma$ you must include the weight of the particle.

Do not use an approximation (i.e. an estimate) for e. Instead, use your work from **a** to replace mg with the tension at the equilibrium level in terms of e.

When you simplify the equation e cancels.

P was raised 15 cm from its equilibrium level.

$x = 0$ at B.

This is also the maximum speed of P.

At the natural length $x = -e$
Use at least 4 s.f. in your approximation for e.

Example (15)   **SKILLS** INTERPRETATION

A particle P of mass $0.2\,\text{kg}$ is attached to one end of a light elastic string of natural length $0.6\,\text{m}$ and modulus of elasticity $8\,\text{N}$. The other end of the string is fixed to a point A on a ceiling. When the particle is hanging in equilibrium the length of the string is $L\,\text{m}$.

a Calculate the value of L.

The particle is held at A and released from rest. It first comes to instantaneous rest when the length of the string is $K\,\text{m}$.

b Use energy considerations to calculate the value of K.

c Show that while the string is taut, P is moving with simple harmonic motion.

The string becomes slack again for the first time T seconds after P was released from A.

d Calculate the value of T.

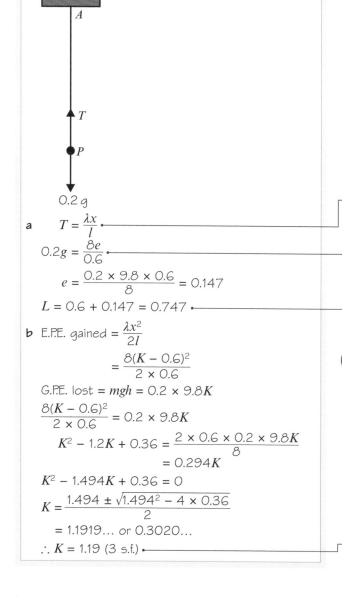

Use Hooke's law.

a $T = \dfrac{\lambda x}{l}$

$0.2g = \dfrac{8e}{0.6}$

At the equilibrium position the tension must equal the weight.

$e = \dfrac{0.2 \times 9.8 \times 0.6}{8} = 0.147$

$L = 0.6 + 0.147 = 0.747$

The total length of the string is required.

b E.P.E. gained $= \dfrac{\lambda x^2}{2l}$

$\qquad = \dfrac{8(K - 0.6)^2}{2 \times 0.6}$

G.P.E. lost $= mgh = 0.2 \times 9.8K$

$\dfrac{8(K - 0.6)^2}{2 \times 0.6} = 0.2 \times 9.8K$

$K^2 - 1.2K + 0.36 = \dfrac{2 \times 0.6 \times 0.2 \times 9.8K}{8}$

$\qquad\qquad\qquad = 0.294K$

$K^2 - 1.494K + 0.36 = 0$

$K = \dfrac{1.494 \pm \sqrt{1.494^2 - 4 \times 0.36}}{2}$

$\quad = 1.1919\ldots \text{ or } 0.3020\ldots$

$\therefore K = 1.19 \text{ (3 s.f.)}$

Problem-solving

The question states that you must do this part using conservation of energy. The kinetic energy is zero at both points under consideration, so the elastic potential energy gained is equal to the gravitational potential energy lost.   ← **Mechanics 3 Section 2.3**

The value of K must be greater than the natural length of the string.

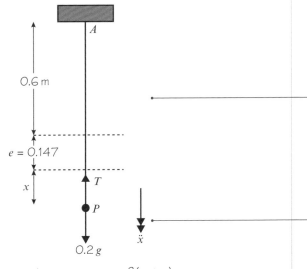

c

0.6 m

$e = 0.147$

x

T

P

$\ddot{x}$

$0.2\,g$

Draw a diagram which shows the natural length and the equilibrium level as well as the distance of P from the centre of the oscillation (x).

Remember that $\ddot{x}$ must be in the direction of increasing x.

$$T = \frac{\lambda \times \text{extension}}{l} = \frac{8(x + e)}{0.6}$$

$$F = ma$$

$$0.2g - \frac{8(x + e)}{0.6} = 0.2\ddot{x}$$

$$\frac{8e}{0.6} - \frac{8(x + e)}{0.6} = 0.2\ddot{x}$$

$$\ddot{x} = -\frac{8x}{0.6 \times 0.2} = -\frac{800}{12}x$$

$\therefore$ S.H.M.

There is no need to use an approximation for e as $\frac{8e}{0.6}$ from part **a**.

Reduce the equation of motion to the form $\ddot{x} = -\omega^2 x$ to establish S.H.M.

d Time to fall 0.6 m from rest

$$s = ut + \frac{1}{2}at^2$$

$$0.6 = 0 + \frac{1}{2} \times 9.8t^2$$

$$t^2 = \frac{0.6}{4.9}$$

$$t = 0.3499\ldots$$

For S.H.M.

$$\omega = \sqrt{\frac{800}{12}} = \sqrt{100 \times \frac{8}{12}} = 10\sqrt{\frac{2}{3}} = \frac{10\sqrt{6}}{3}$$

amplitude $= K - L = 1.191 - 0.747 = 0.444$

$x = 0.444\cos\omega t$

when $x = 0.147$, $0.147 = 0.444\cos\omega t$

$$\cos\omega t = \frac{0.147}{0.444}$$

$$t = \frac{1}{\omega}\arccos\left(\frac{0.147}{0.444}\right) = 0.1510\ldots$$

Period $= \dfrac{2\pi}{\omega} = 2\pi \times \dfrac{3}{10\sqrt{6}} = 0.7695\ldots$

Time for which string is taut

$= 0.7695 - 2 \times 0.1510 = 0.4675\ldots$

Total time $= 0.4675\ldots + 0.3499 = 0.8174\ldots$

$\therefore T = 0.817$ (3 s.f.)

Until the string is taut, P is falling freely under gravity.

Problem-solving

Because of the symmetry of S.H.M. there are several ways to obtain the time for which the string is taut. Whichever method you use you must show your working clearly.

Using $x = a\cos\omega t$ with the positive value of x when the string is at its natural length will give the time from the high point of the oscillation (if it were complete) to the point where the string becomes taut.

Subtracting twice the time just found from the period will give the time for which the string is taut in any one oscillation.

Finally, add the time taken while falling freely under gravity to the time for which the string is taut.

Exercise **3E** **SKILLS** **INTERPRETATION**

Whenever a numerical value of g is required, take $g = 9.8\,\text{m s}^{-2}$

E/P **1** A particle P of mass $0.75\,\text{kg}$ is hanging in equilibrium attached to one end of a light elastic spring of natural length $1.5\,\text{m}$ and modulus of elasticity $80\,\text{N}$. The other end of the spring is attached to a fixed point A vertically above P.

    a Calculate the length of the spring. **(3 marks)**

    The particle is pulled downwards and held at a point B which is vertically below A. The particle is then released from rest.

    b Show that P moves with simple harmonic motion. **(4 marks)**

    c Calculate the period of the oscillations. **(2 marks)**

    The particle passes through its equilibrium position with speed $2.5\,\text{m s}^{-1}$.

    d Calculate the amplitude of the oscillations. **(4 marks)**

E/P **2** A particle P of mass $0.5\,\text{kg}$ is attached to the free end of a light elastic spring of natural length $0.5\,\text{m}$ and modulus of elasticity $50\,\text{N}$. The other end of the spring is attached to a fixed point A and P hangs in equilibrium vertically below A.

    a Calculate the extension of the spring. **(3 marks)**

    The particle is now pulled vertically down a further $0.2\,\text{m}$ and released from rest.

    b Calculate the period of the resulting oscillations. **(2 marks)**

    c Calculate the maximum speed of the particle. **(2 marks)**

3 A particle P of mass $2\,\text{kg}$ is hanging in equilibrium attached to the free end of a light elastic spring of natural length $1.5\,\text{m}$ and modulus of elasticity $\lambda\,\text{N}$. The other end of the spring is fixed to a point A vertically above P. The particle receives an impulse of magnitude $3\,\text{N s}$ in the direction AP.

    a Find the speed of P immediately after the impact.

    b Show that P moves with simple harmonic motion.

    The period of the oscillations is $\frac{\pi}{2}\,\text{s}$.

    c Find the value of λ.

    d Find the amplitude of the oscillations.

E/P **4** A piston of mass $2\,\text{kg}$ moves inside a smooth cylinder which is fixed with its axis vertical. The piston is attached to the base of the cylinder by a spring of natural length $12\,\text{cm}$ and modulus of elasticity $500\,\text{N}$. The piston is released from rest at a point where the spring is compressed to a length of $8\,\text{cm}$. Assuming that the spring can be modelled as a light elastic spring and the piston as a particle, calculate

    a the period of the resulting oscillations **(5 marks)**

    b the maximum speed of the piston. **(2 marks)**

(E/P) **5** A light elastic string of natural length 40 cm has one end A attached to a fixed point. A particle P of mass 0.4 kg is attached to the free end of the string and hangs freely in equilibrium vertically below A. The distance AP is 45 cm.

    a Find the modulus of elasticity of the string. **(3 marks)**

    The particle is now pulled vertically downwards until AP measures 52 cm and then released from rest.

    b Show that, while the string is taut, P moves with simple harmonic motion. **(4 marks)**

    c Find the period and amplitude of the motion. **(3 marks)**

    d Find the greatest speed of P during the motion. **(2 marks)**

    e Find the time taken by P to rise 11 cm from the point of release. **(3 marks)**

(E/P) **6** A particle P of mass 0.4 kg is attached to one end of a light elastic string of natural length 0.5 m and modulus of elasticity 10 N. The other end of the string is attached to a fixed point A and P is initially hanging freely in equilibrium vertically below A. The particle is then pulled vertically downwards a further 0.2 m and released from rest.

    a Calculate the time from release until the string becomes slack for the first time. **(4 marks)**

    b Calculate the time between the string first becoming slack and the next time it becomes taut. **(4 marks)**

(E/P) **7** A particle P of mass 1.5 kg is hanging freely attached to one end of a light elastic string of natural length 1 m and modulus of elasticity 40 N. The other end of the string is attached to a fixed point A on a ceiling. The particle is pulled vertically downwards until AP is 1.8 m and released from rest. When P has risen a distance 0.4 m the string is cut.

    a Calculate the greatest height P reaches above its equilibrium position. **(4 marks)**

    b Calculate the time taken from release to reach that greatest height. **(3 marks)**

(E/P) **8** A particle P of mass 1.5 kg is attached to the midpoint of a light elastic string of natural length 1.2 m and modulus of elasticity 15 N. The ends of the string are fixed to the points A and B where A is vertically above B and $AB = 2.8$ m

    a Given that P is in equilibrium calculate the length AP. **(3 marks)**

    The particle is now pulled downwards a distance 0.15 m from its equilibrium position and released from rest.

    b Prove that P moves with simple harmonic motion. **(4 marks)**

    T seconds after being released P is 0.1 m above its equilibrium position.

    c Find the value of T. **(3 marks)**

(E/P) **9** A rock climber of mass 70 kg is attached to one end of a rope. He falls from a place which is 8 m vertically below the point to which the other end of the rope is fixed. The climber falls vertically without hitting the rock face. Assuming that the climber can be modelled as a particle and the rope as a light elastic string of natural length 16 m and modulus of elasticity 40 000 N, calculate

    a the climber's speed at the instant when the rope becomes taut **(3 marks)**

    b the maximum distance of the climber below the ledge **(3 marks)**

    c the time from falling from the ledge to reaching his lowest point. **(2 marks)**

Challenge

A particle P of mass m kg is attached to one end of a light elastic string of natural length l m and modulus of elasticity $5mg$. the other end of the string is attached to a fixed point A on a ceiling. The particle is pulled vertically downwards and released to oscillate with period T s.

A second particle Q of mass km kg is then also attached to the end of the string. The system then oscillates with period $3T$ s.

Find the value of k.

Chapter review (3)   **SKILLS** ▷ PROBLEM-SOLVING

Whenever a numerical value of g is required, take $g = 9.8 \, \text{ms}^{-2}$

(E/P) **1** A particle P of mass 0.6 kg moves along the positive x-axis under the action of a single force which is directed towards the origin O and has magnitude $\dfrac{k}{(x+2)^2}$ N where $OP = x$ metres and k is a constant. Initially P is moving away from O. At $x = 2$ the speed of P is $8 \, \text{ms}^{-1}$ and at $x = 10$ the speed of P is $2 \, \text{ms}^{-1}$.

   a Find the value of k.   **(6 marks)**

   The particle first comes to instantaneous rest at the point B.

   b Find the distance OB.   **(4 marks)**

(E) **2** A particle P of mass 0.5 kg is moving along the x-axis, in the positive x direction. At time t seconds (where $t > 0$) the resultant force acting on P has magnitude $\dfrac{5}{\sqrt{(3t+4)}}$ N and is directed towards the origin O. When $t = 0$, P is moving through O with speed $12 \, \text{ms}^{-1}$.

   a Find an expression for the velocity of P at time t seconds.   **(5 marks)**

   b Find the distance of P from O when P is instantaneously at rest.   **(5 marks)**

(E/P) **3** A spacecraft S of mass m is moving in a straight line towards the centre of the Earth. When the distance of S from the centre of the Earth is x metres, the force exerted by the Earth on S has magnitude $\dfrac{k}{x^2}$, where k is a constant, and is directed towards the centre of the Earth.

   a By modelling the Earth as a sphere of radius R and S as a particle, show that $k = mgR^2$

                                                   (2 marks)

   The spacecraft starts from rest when $x = 5R$

   b Assuming that air resistance can be ignored, find the speed of S as it crashes onto the Earth's surface.   **(7 marks)**

(E) **4** A particle P is moving with simple harmonic motion between two points A and B which are 0.4 m apart on a horizontal line. The midpoint of AB is O. At time $t = 0$, P passes through O, moving towards A, with speed $u \, \text{ms}^{-1}$. The next time P passes through O is when $t = 2.5$ s

   a Find the value of u.   **(4 marks)**

   b Find the speed of P when $t = 3$ s   **(2 marks)**

   c Find the distance of P from A when $t = 3$ s   **(5 marks)**

(E) **5** A particle P of mass $1.2\,\text{kg}$ moves along the x-axis. At time $t = 0$, P passes through the origin O, moving in the positive x direction. At time t seconds, the velocity of P is $v\,\text{m s}^{-1}$ and $OP = x$ metres. The resultant force acting on P has magnitude $6(2.5 - x)\,\text{N}$ and acts in the positive x direction. The maximum speed of P is $8\,\text{m s}^{-1}$.

    a Find the value of x when the speed of P is $8\,\text{m s}^{-1}$. **(5 marks)**

    b Find an expression for v^2 in terms of x. **(5 marks)**

(E/P) **6** A particle P moves along the x-axis in such a way that at time t seconds its distance x metres from the origin O is given by $x = 3\sin\left(\dfrac{\pi t}{4}\right)$

    a Prove that P moves with simple harmonic motion. **(4 marks)**

    b Write down the amplitude and the period of the motion. **(3 marks)**

    c Find the maximum speed of P. **(2 marks)**

    The points A and B are on the same side of O with $OA = 1.2\,\text{m}$ and $OB = 2\,\text{m}$.

    d Find the time taken by P to travel directly from A to B. **(4 marks)**

(E/P) **7** A particle P moves on the x-axis with simple harmonic motion such that its centre of oscillation is the origin, O. When P is a distance $0.09\,\text{m}$ from O, its speed is $0.3\,\text{m s}^{-1}$ and the magnitude of its acceleration is $1.5\,\text{m s}^{-2}$

    a Find the period of the motion. **(3 marks)**

    The amplitude of the motion is a metres. Find

    b the value of a **(3 marks)**

    c the total time, within one complete oscillation, for which the distance OP is greater than $\dfrac{a}{2}$ metres. **(5 marks)**

(E/P) **8** A particle P of mass $0.6\,\text{kg}$ is attached to one end of a light elastic spring of natural length $2.5\,\text{m}$ and modulus of elasticity $25\,\text{N}$. The other end of the spring is attached to a fixed point A on the smooth horizontal table on which P lies. The particle is held at the point B where $AB = 4\,\text{m}$ and released from rest.

    a Prove that P moves with simple harmonic motion. **(4 marks)**

    b Find the period and amplitude of the motion. **(3 marks)**

    c Find the time taken for P to move $2\,\text{m}$ from B. **(2 marks)**

(E/P) **9** A particle P of mass $0.4\,\text{kg}$ is attached to the midpoint of a light elastic string of natural length $1.2\,\text{m}$ and modulus of elasticity $2.5\,\text{N}$. The ends of the string are attached to points A and B on a smooth horizontal table where $AB = 2\,\text{m}$. The particle P is released from rest at the point C on the table, where A, C and B lie in a straight line and $AC = 0.7\,\text{m}$

    a Show that P moves with simple harmonic motion. **(4 marks)**

    b Find the period of the motion. **(3 marks)**

    The point D lies between A and B and $AD = 0.85\,\text{m}$

    c Find the time taken by P to reach D for the first time. **(4 marks)**

(E/P) **10** A and B are two points on a smooth horizontal floor, where $AB = 12\,\text{m}$

A particle P has mass $0.4\,\text{kg}$. One end of a light elastic spring, of natural length $5\,\text{m}$ and modulus of elasticity $20\,\text{N}$, is attached to P and the other end is attached to A. The ends of another light elastic spring, of natural length $3\,\text{m}$ and modulus of elasticity $18\,\text{N}$, are attached to P and B.

a Find the extensions in the two springs when the particle is at rest in equilibrium. **(5 marks)**

Initially P is at rest in equilibrium. It is then set in motion and starts to move towards B. In the subsequent motion P does not reach A or B.

b Show that P oscillates with simple harmonic motion about the equilibrium position.

(4 marks)

c Given that P stays within $0.4\,\text{m}$ of the equilibrium position for $\dfrac{1}{3}$ of the time within each complete oscillation, find the initial speed of P. **(7 marks)**

(E/P) **11** A particle P of mass $0.5\,\text{kg}$ is attached to one end of a light elastic string of natural length $1.2\,\text{m}$ and modulus of elasticity $\lambda\,\text{N}$. The other end of the string is attached to a fixed point A. The particle is hanging in equilibrium at the point O, which is $1.4\,\text{m}$ vertically below A.

a Find the value of λ. **(3 marks)**

The particle is now displaced to a point B, $1.75\,\text{m}$ vertically below A, and released from rest.

b Prove that while the string is taut P moves with simple harmonic motion. **(4 marks)**

c Find the period of the simple harmonic motion. **(4 marks)**

d Calculate the speed of P at the first instant when the string becomes slack. **(4 marks)**

e Find the greatest height reached by P above O. **(4 marks)**

(E/P) **12** A particle P of mass m is attached to the midpoint of a light elastic string of natural length $4l$ and modulus of elasticity $5mg$. One end of the string is attached to a fixed point A and the other end to a fixed point B, where A and B lie on a smooth horizontal surface and $AB = 6l$. The particle is held at the point C, where A, C and B are **collinear** and $AC = \dfrac{9l}{4}$, and released from rest.

a Prove that P moves with simple harmonic motion. **(4 marks)**

Find, in terms of g and l

b the period of the motion **(2 marks)**

c the maximum speed of P. **(2 marks)**

Challenge

The motion of a space shuttle which is launched from a point O on the surface of the Earth can be modelled as a particle of mass m moving in a straight line, subject to the universal law of gravitation using $F = -\dfrac{mMG}{(R + x)^2}$ where

M is the mass of the Earth

m is the mass of the space shuttle

R is the radius of the Earth

x is the height of the space shuttle above the Earth

G is the universal constant of gravitation.

a Given a space shuttle is launched with initial velocity u m s^{-1}, show that the maximum height, H, above the Earth that the spaceship reaches can be expressed as $H = \dfrac{Ru^2}{\left(\dfrac{2MG}{R} - u^2\right)}$

The minimum velocity required to project the space shuttle into space is called the **escape velocity**. This is the value of u for which H tends to infinity.

b Use

$M = 5.98 \times 10^{24}$

$R = 6.4 \times 10^{6}$

$G = 6.7 \times 10^{-11}$

to work out the escape velocity for the space shuttle correct to 3 significant figures.

Summary of key points

1   In forming an equation of motion, forces that tend to decrease the displacement are negative and forces that tend to increase the displacement are positive.

2   Newton's law of gravitation states that the force of attraction between two bodies of mass M_1 and M_2 is directly proportional to the product of their masses and inversely proportional to the square of the distance between them.

   • $F = \dfrac{GM_1M_2}{d^2}$ where G is a constant known as the constant of gravitation.

3   Simple harmonic motion (S.H.M.) is motion in which the acceleration of a particle is always towards a fixed point O on the line of motion of P, and has magnitude proportional to the displacement of P from O.

4 For S.H.M. of amplitude a defined by the equation $\ddot{x} = -\omega^2 x$
  - $v^2 = \omega^2(a^2 - x^2)$
  - If P is at the centre of the oscillation when $t = 0$, use $x = a \sin \omega t$
  - If P is at an end point of the oscillation when $t = 0$, use $x = a \cos \omega t$
  - If P is at some other point when $t = 0$, use $x = a \sin(\omega t + \alpha)$
  - The period of the oscillation is $T = \dfrac{2\pi}{\omega}$

5 For a particle moving on a smooth horizontal surface attached to one end of an elastic spring
  - the particle will move with S.H.M.
  - the particle will perform complete oscillations.

6 For a particle moving on a smooth horizontal surface attached to one end of an elastic string
  - the particle will move with S.H.M. while the string is taut
  - the particle will move with constant speed while the string is slack.

7 To solve problems involving elastic springs and strings
  - use Hooke's law to find the tension
  - use $F = ma$ to obtain ω
  - use information given in the question to obtain the amplitude.

8 For a particle hanging in equilibrium attached to one end of an elastic spring and displaced vertically from its equilibrium position
  - the particle will move with S.H.M.
  - the particle will perform complete oscillations
  - the centre of the oscillation will be the equilibrium position.

9 For a particle hanging in equilibrium attached to one end of an elastic string and displaced vertically from its equilibrium position
  - the particle will move with S.H.M. while the string is taut
  - the particle will perform complete oscillations if the amplitude is no greater than the equilibrium extension
  - if the amplitude is greater than the equilibrium extension the particle will move freely under gravity while the string is slack.

Review exercise

1

1 A particle P moves in a straight line. At time t seconds, the acceleration of P is $e^{2t} \, \text{m s}^{-2}$, where $t \geqslant 0$. When $t = 0$, P is at rest. Show that the speed, $v \, \text{m s}^{-1}$, of P at time t seconds is given by

$$v = \frac{1}{2}(e^{2t} - 1) \qquad \text{(6)}$$

← Mechanics 3 Section 1.1

2 A particle P moves along the x-axis in the positive direction. At time t seconds, the velocity of P is $v \, \text{m s}^{-1}$ and its acceleration is $\frac{1}{2}e^{-\frac{1}{6}t} \, \text{m s}^{-2}$. When $t = 0$ the speed of P is $10 \, \text{m s}^{-1}$.

  a Express v in terms of t. **(6)**

  b Find, to 3 significant figures, the speed of P when $t = 3$. **(2)**

  c Find the limiting value of v. **(1)**

← Mechanics 3 Section 1.1

3 A particle P moves along the x-axis. At time t seconds the velocity of P is $v \, \text{m s}^{-1}$ and its acceleration is $2 \sin \frac{1}{2}t \, \text{m s}^{-2}$, both measured in the direction Ox. Given that $v = 4$ when $t = 0$,

  a find v in terms of t **(6)**

  b calculate the distance travelled by P between the times $t = 0$ and $t = \frac{\pi}{2}$ **(7)**

← Mechanics 3 Section 1.1

(E/P) 4 At time $t = 0$, a particle P is at the origin O and is moving with speed $18 \, \text{m s}^{-1}$ along the x-axis in the positive x direction. At time t seconds ($t > 0$), the acceleration of P has magnitude $\dfrac{3}{\sqrt{t+4}} \, \text{m s}^{-2}$ and is directed towards O.

  a Show that, at time t seconds, the velocity of P is $(30 - 6\sqrt{t+4}) \, \text{m s}^{-1}$. **(6)**

  b Find the distance of P from O when P comes to instantaneous rest. **(7)**

← Mechanics 3 Section 1.1

(E/P) 5 A particle moving in a straight line starts from rest at a point O at time $t = 0$. At time t seconds, the velocity $v \, \text{m s}^{-1}$ is given by

$$v = \begin{cases} 3t(t - 4), & 0 \leqslant t \leqslant 5 \\ 75t^{-1}, & 5 < t \leqslant 10 \end{cases}$$

  a Sketch a velocity–time graph for the particle for $0 \leqslant t \leqslant 10$ **(3)**

  b Find the set of values of t for which the acceleration of the particle is positive. **(2)**

  c Show that the total distance travelled by the particle in the interval $0 \leqslant t \leqslant 5$ is 39 m. **(5)**

  d Find, to 3 significant figures, the value of t at which the particle returns to O. **(3)**

← Mechanics 3 Section 1.1

(E) 6 A particle P moves along the x-axis in such a way that when its displacement from the origin O is x m, its velocity is $v \, \text{m s}^{-1}$ and its acceleration is $4x \, \text{m s}^{-2}$. When $x = 2$, $v = 4$

Show that $v^2 = 4x^2$ **(4)**

← Mechanics 3 Section 1.2

7 A particle P moves on the positive x-axis. When $OP = x$ metres, where O is the origin, the acceleration of P is directed away from O and has magnitude $\left(1 - \dfrac{4}{x^2}\right)$ m s^{-2}. When $OP = x$ metres, the velocity of P is v m s^{-1}. Given that when $x = 1$, $v = 3\sqrt{2}$, show that when

$$x = \frac{3}{2}, \quad v^2 = \frac{49}{3} \hspace{2cm} \textbf{(5)}$$

← Mechanics 3 Section 1.2

8 A particle P is moving in a straight line. When P is at a distance x metres from a fixed point O on the line, the acceleration of P is $(5 + 3\sin 3x)$ m s^{-2} in the direction OP. Given that P passes through O with speed 4 m s^{-1}, find the speed of P at $x = 6$ Give your answer to 3 significant figures. **(5)**

← Mechanics 3 Section 1.2

9 A particle P is moving along the positive x-axis in the direction of x increasing. When $OP = x$ metres, the velocity of P is v m s^{-1} and the acceleration of P is $\dfrac{4k^2}{(x+1)^2}$ m s^{-2}, where k is a positive constant. At $x = 1$, $v = 0$.

a Find v^2 in terms of x and k. **(5)**

b Deduce that v cannot exceed $2k$. **(2)**

← Mechanics 3 Section 1.2

10 A particle P moves along the x-axis. At time $t = 0$, P passes through the origin O, moving in the positive x direction. At time t seconds, the velocity of P is v m s^{-1} and $OP = x$ metres. The acceleration of P is $\dfrac{1}{12}(30 - x)$ m s^{-2}, measured in the positive x direction.

a Give a reason why the maximum speed of P occurs when $x = 30$ **(2)**

Given that the maximum speed of P is 10 m s^{-1},

b find an expression for v^2 in terms of x. **(5)**

← Mechanics 3 Section 1.2

11 A particle P starts at rest and moves in a straight line. The acceleration of P initially has magnitude 20 m s^{-2} and, in a first model of the motion of P, it is assumed that this acceleration remains constant.

a For this model, find the distance moved by P while accelerating from rest to a speed of 6 m s^{-1}. **(4)**

The acceleration of P when it is x metres from its initial position is a m s^{-2} and it is then established that $a = 12$ when $x = 2$ A refined model is proposed in which $a = p - qx$, where p and q are constants.

b Show that, under the refined model, $p = 20$ and $q = 4$ **(5)**

c Hence find, for the refinded model, the distance moved by P in first attaining a speed of 6 m s^{-1}. **(4)**

← Mechanics 3 Section 1.2

12 A particle P lies in equilibrium on a horizontal smooth surface, and is attached to points A and B on the same surface by means of two springs. Spring AP has natural length 0.8 m and modulus of elasticity 24 N, and spring PB has natural length 0.4 m and modulus of elasticity 20 N. Given that APB is a straight line and that the distance AB is 1.6 m, find

a the distance AP **(6)**

b the tension in each spring. **(2)**

← Mechanics 3 Section 2.1

13 The elastic springs AB and BC are joined at B to form one long spring. The ends of the long spring are attached to two fixed points 4 m apart. The spring AB has natural length 1.5 m and modulus of elasticity 20 N, and the spring BC has natural length 0.75 m and modulus of elasticity 15 N.

Find the lengths AB and BC. **(8)**

← Mechanics 3 Section 2.1

E **14**

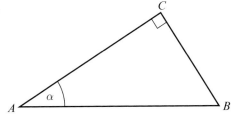

A rod AB, of mass $2m$ and length $2a$, is suspended from a fixed point C by two light strings AC and BC. The rod rests horizontally in equilibrium with AC making an angle α with the rod, where $\tan \alpha = \frac{3}{4}$, and with AC perpendicular to BC, as shown in the figure.

a Give a reason why the rod cannot be uniform. **(1)**

b Show that the tension in BC is $\frac{8}{5}mg$ and find the tension in AC. **(5)**

The string BC is elastic, with natural length a and modulus of elasticity kmg, where k is a constant.

c Find the value of k. **(4)**

← Mechanics 3 Section 2.1

E **15**

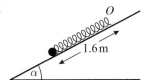

A particle of mass $0.8\,\text{kg}$ is attached to one end of a light elastic spring, of natural length $2\,\text{m}$ and modulus of elasticity $20\,\text{N}$. The other end of the spring is attached to a fixed point O on a smooth plane which is inclined at an angle α to the horizontal, where $\tan \alpha = \frac{3}{4}$. The particle is held at a point which is $1.6\,\text{m}$ down the line of greatest slope of the plane from O, as shown in the figure. The particle is then released from rest.

Find the initial acceleration of the particle. **(6)**

← Mechanics 3 Section 2.2

E/P **16**

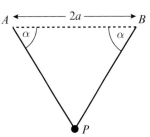

Two light elastic strings each have natural length a and modulus of elasticity λ. A particle P of mass m is attached to one end of each string. The other ends of the string are attached to points A and B, where AB is horizontal and $AB = 2a$ The particle is held at the midpoint of AB and released from rest. It comes to rest for the first time in its subsequent motion when PA and PB make angles α with AB, where $\tan \alpha = \frac{4}{3}$, as shown in the figure. Find λ in terms of m and g. **(7)**

← Mechanics 3 Section 2.2

E/P **17** A light elastic string AB of natural length $1.5\,\text{m}$ has modulus of elasticity $20\,\text{N}$. The end A is fixed to a point on a smooth horizontal table. A small ball S of mass $0.2\,\text{kg}$ is attached to the end B. Initially S is at rest on the table with $AB = 1.5\,\text{m}$ The ball S is then projected horizontally directly away from A with a speed of $5\,\text{m\,s}^{-1}$. By modelling S as a particle,

a find the speed of S when $AS = 2\,\text{m}$ **(5)**

When the speed of S is $1.5\,\text{m\,s}^{-1}$, the string breaks.

b Find the tension in the string immediately it breaks. **(5)**

← Mechanics 3 Section 2.2

E/P **18** One end of an elastic string of natural length $l\,\text{m}$ and modulus of elasticity $\lambda\,\text{N}$ is fixed to a ceiling. A particle of mass $m\,\text{kg}$ is attached to the free end and hangs in equilibrium. Show that the elastic potential energy stored by the string is given by $\dfrac{m^2 g^2 l}{2\lambda}$ **(5)**

← Mechanics 3 Section 2.3

E/P 19 An elastic string has natural length 0.5 m and modulus of elasticity 20 N. One end of the string is fixed. A particle of mass 0.5 kg is attached to the free end and hangs in equilibrium. The string is then stretched to a length of 1 m.

Calculate the work done in stretching the string.   **(6)**

← Mechanics 3 Section 2.3

E 20

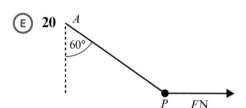

A particle of mass 0.8 kg is attached to one end of a light elastic string, of natural length 1.2 m and modulus of elasticity 24 N. The other end of the string is attached to a fixed point A. A horizontal force of magnitude F newtons is applied to P. The particle is in equilibrium with the string making an angle 60° with the downward vertical as shown in the figure. Calculate

a the value of F **(3)**

b the extension of the string   **(3)**

c the elastic energy stored in the string.   **(2)**

← Mechanics 3 Sections 2.3, 2.4

E/P 21 A particle P of mass m is attached to one end of a light elastic string, of natural length a and modulus of elasticity $3.6mg$. The other end of the string is fixed at a point O on a rough horizontal table. The particle is projected along the surface of the table from O with speed $\sqrt{2ag}$

At its furthest point from O, the particle is at the point A, where $OA = \dfrac{4}{3}a$

a Find, in terms of m, g and a, the elastic energy stored in the string when P is at A.   **(3)**

b Using the work–energy principle, or otherwise, find the coefficient of friction between P and the table.   **(6)**

← Mechanics 3 Sections 2.3, 2.4

E/P 22 A particle P of mass m is held at a point A on a rough horizontal plane. The coefficient of friction between P and the plane is $\dfrac{2}{3}$. The particle is attached to one end of a light elastic string, of natural length a and modulus of elasticity $4mg$. The other end of the string is attached to a fixed point O on the plane, where

$$OA = \dfrac{3}{2}a$$

The particle P is released from rest and comes to rest at a point B, where $OB < a$

Using the work–energy principle, or otherwise, calculate the distance AB.   **(6)**

← Mechanics 3 Sections 2.3, 2.4

E/P 23 A particle of mass 5 kg is attached to one end of a spring of natural length 1 m and modulus of elasticity 75 N. The other end of the spring is fixed to a point, P, on a smooth horizontal table. The particle is held 1.5 m from P and then released.

Show that the speed of the particle when the spring reaches its natural length is

$$\dfrac{\sqrt{15}}{2}\,\text{m s}^{-1}$$ **(4)**

← Mechanics 3 Section 2.4

E/P 24 A light elastic string, of natural length 0.8 m and modulus of elasticity 15 N has one end attached to a fixed point P. A particle of mass 0.5 kg is attached to the other end. The particle is held at a point which is 2 m vertically below P and released from rest.

Find the speed of the particle

a when the string first becomes slack   **(4)**

b when the particle reaches P.   **(2)**

← Mechanics 3 Section 2.4

(E/P) 25 A light elastic string has natural length 4 m and modulus of elasticity 58.8 N. A particle P of mass 0.5 kg is attached to one end of the string. The other end of the string is attached to a fixed point A. The particle is released from rest at A and falls vertically.

  a Find the distance travelled by P before it comes to instantaneous rest for the first time. **(7)**

  The particle is now held at a point 7 m vertically below A and released from rest.

  b Find the speed of the particle when the string first becomes slack. **(5)**

← Mechanics 3 Section 2.4

(E) 26 A particle P of mass 2.5 kg moves along the positive x-axis. It moves away from a fixed origin O, under the action of a force directed away from O. When $OP = x$ metres, the magnitude of the force is $2e^{-0.1x}$ N and the speed of P is v m s^{-1}. When $x = 0$, $v = 2$. Find

  a v^2 in terms of x **(4)**

  b the value of x when $v = 4$. **(2)**

  c Give a reason why the speed of P does not exceed $\sqrt{20}$ m s^{-1}. **(2)**

← Mechanics 3 Section 3.1

(E/P) 27 A toy car of mass 0.2 kg is travelling in a straight line on a horizontal floor. The car is modelled as a particle. At time $t = 0$ the car passes through a fixed point O. After t seconds the speed of the car is v m s^{-1} and the car is at a point P with $OP = x$ metres. The resultant force on the car is modelled as $\frac{1}{10}x(4 - 3x)$ N in the direction OP. The car comes to instantaneous rest when $x = 6$. Find

  a an expression for v^2 in terms of x **(4)**

  b the initial speed of the car. **(2)**

← Mechanics 3 Section 3.1

(E) 28 A car of mass 800 kg moves along a horizontal straight road. At time t seconds, the resultant force on the car has magnitude $\dfrac{48\,000}{(t + 2)^2}$ N, acting in the direction of motion of the car. When $t = 0$, the car is at rest.

  a Show that the speed of the car approaches a limiting value as t increases and find this value. **(7)**

  b Find the distance moved by the car in the first 6 s of its motion. **(5)**

← Mechanics 3 Section 3.1

(E/P) 29 A particle P of mass $\frac{1}{3}$ kg moves along the positive x-axis under the action of a single force. The force is directed towards the origin O and has magnitude $\dfrac{k}{(x + 1)^2}$ N, where $OP = x$ metres and k is a constant. Initially P is moving away from O. At $x = 1$ the speed of P is 4 m s^{-1}, and at $x = 8$ the speed of P is $\sqrt{2}$ m s^{-1}.

Find

  a the value of k **(7)**

  b the distance of P from O when P first comes to instantaneous rest. **(5)**

← Mechanics 3 Section 3.1

(E/P) 30 Above the Earth's surface, the magnitude of the force on a particle due to the Earth's gravity is inversely proportional to the square of the distance of the particle from the centre of the Earth. Assuming that the Earth is a sphere of radius R, and taking g as the acceleration due to gravity at the surface of the Earth,

  a prove that the magnitude of the gravitational force on a particle of mass m when it is a distance x (where $x \geq R$) from the centre of the Earth is $\dfrac{mgR^2}{x^2}$ **(3)**

A particle is fired vertically upwards from the surface of the Earth with initial speed u, where $u^2 = \dfrac{3}{2}gR$

Ignoring air resistance,

b find, in terms of g and R, the speed of the particle when it is at a height $2R$ above the surface of the Earth. **(7)**

← Mechanics 3 Section 3.2

(E/P) **31** A projectile P is fired vertically upwards from a point on the Earth's surface. When P is at a distance x from the centre of the Earth its speed is v. Its acceleration is directed towards the centre of the Earth and has magnitude $\dfrac{k}{x^2}$ where k is a constant. The Earth is assumed to be a sphere of radius R.

a Show that the motion of P may be modelled by the differential equation

$$v\frac{dv}{dx} = -\frac{gR^2}{x^2} \qquad \textbf{(3)}$$

The initial speed of P is U, where $U^2 < 2gR$. The greatest distance of P from the centre of the Earth is X.

b Find X in terms of U, R and g. **(7)**

← Mechanics 3 Section 3.2

(E) **32** A particle P moves in a straight line with simple harmonic motion about a fixed centre O with period 2 s. At time t seconds the speed of P is $v\,\text{m s}^{-1}$. When $t = 0$, $v = 0$ and P is at a point A where $OA = 0.25\,\text{m}$

Find the smallest positive value of t for which $AP = 0.375\,\text{m}$ **(5)**

← Mechanics 3 Section 3.3

(E/P) **33** A piston P in a machine moves in a straight line with simple harmonic motion about a fixed centre O. The period of the oscillations is π s. When P is 0.5 m from O, its speed is $2.4\,\text{m s}^{-1}$. Find

a the amplitude of the motion **(4)**

b the maximum speed of P during its motion **(2)**

c the maximum magnitude of the acceleration of P during the motion **(2)**

d the total time, in seconds to 2 decimal places, in each complete oscillation for which the speed of P is greater than $2.4\,\text{m s}^{-1}$. **(5)**

← Mechanics 3 Section 3.3

(E) **34** The points O, A, B and C lie in a straight line, in that order, with $OA = 0.6\,\text{m}$, $OB = 0.8\,\text{m}$ and $OC = 1.2\,\text{m}$. A particle P, moving in a straight line, has speed $\left(\dfrac{3}{10}\sqrt{3}\right)\text{m s}^{-1}$ at A, $\left(\dfrac{1}{5}\sqrt{5}\right)\text{m s}^{-1}$ at B and is instantaneously at rest at C.

a Show that this information is consistent with P performing simple harmonic motion with centre O. **(4)**

Given that P is performing simple harmonic motion with centre O,

b show that the speed of P at O is $0.6\,\text{m s}^{-1}$. **(2)**

c Find the magnitude of the acceleration of P as it passes A. **(4)**

d Find, to 3 significant figures, the time taken for P to move directly from A to B. **(4)**

← Mechanics 3 Section 3.3

(E/P) **35** A piston in a machine is modelled as a particle of mass 0.2 kg attached to one end A of a light elastic spring, of natural length 0.6 m and modulus of elasticity 48 N. The other end B of the spring is fixed and the piston is free to move in a horizontal tube which is assumed to be smooth. The piston is released from rest when $AB = 0.9\,\text{m}$

a Prove that the motion of the piston is simple harmonic with period $\dfrac{\pi}{10}$ s **(4)**

b Find the maximum speed of the piston. **(2)**

c Find, in terms of π, the length of time during each oscillation for which the length of the spring is less than 0.75 m. **(5)**

← **Mechanics 3 Section 3.4**

(E/P) 36 A particle P of mass 0.8 kg is attached to one end A of a light elastic spring OA, of natural length 60 cm and modulus of elasticity 12 N. The spring is placed on a smooth table and the end O is fixed. The particle is pulled away from O to a point B, where $OB = 85$ cm, and is released from rest.

a Prove that the motion of P is simple harmonic motion with period $\dfrac{2\pi}{5}$ s **(4)**

b Find the greatest magnitude of the acceleration of P during the motion. **(2)**

Two seconds after being released from rest, P passes through the point C.

c Find, to 2 significant figures, the speed of P as it passes through C. **(2)**

d State the direction in which P is moving 2 s after being released. **(2)**

← **Mechanics 3 Section 3.4**

(E/P) 37 A particle P of mass 0.3 kg is attached to one end of a light elastic spring. The other end of the spring is attached to a fixed point O on a smooth horizontal table. The spring has natural length 2 m and modulus of elasticity 21.6 N. The particle P is placed on the table at a point A, where $OA = 2$ m. The particle P is now pulled away from O to the point B, where OAB is a straight line with $OB = 3.5$ m It is then released from rest.

a Prove that P moves with simple harmonic motion of period $\dfrac{\pi}{3}$ s **(4)**

b Find the speed of P when it reaches A. **(2)**

The point C is the midpoint of AB.

c Find, in terms of π, the time taken for P to reach C for the first time. **(4)**

Later in the motion, P collides with a particle Q of mass 0.2 kg which is at rest at A.

After impact, P and Q coalesce to form a single particle R.

d Show that R also moves with simple harmonic motion and find the amplitude of this motion. **(4)**

← **Mechanics 3 Section 3.4**

(E/P) 38 A light elastic string, of natural length $4a$ and modulus of elasticity $8mg$, has one end attached to a fixed point A. A particle P of mass m is attached to the other end of the string and hangs in equilibrium at the point O.

a Find the distance AO. **(4)**

The particle is now pulled down to a point C vertically below O, where $OC = d$ It is released from rest. In the subsequent motion the string does not become slack.

b Show that P moves with simple harmonic motion of period $\pi\sqrt{\dfrac{2a}{g}}$ **(5)**

The greatest speed of P during this motion is $\dfrac{1}{2}\sqrt{ga}$

c Find d in terms of a. **(2)**

Instead of being pulled down a distance d, the particle is pulled down a distance a. Without further calculation,

d describe briefly the subsequent motion of P. **(2)**

← **Mechanics 3 Section 3.5**

Challenge

1 A particle P travels on the x-axis, passing the origin at time $t = 0$ with velocity $-k\,\mathrm{m\,s^{-1}}$ where k is a positive constant. At time t the particle is a distance $x\,\mathrm{m}$ from the origin and its acceleration, $a\,\mathrm{m\,s^{-2}}$, is given by $a = 8x\dfrac{\mathrm{d}x}{\mathrm{d}t}$

Show that the distance of the particle from the origin never exceeds $\frac{1}{2}\sqrt{k}$ metres.

← **Mechanics 3 Section 1.2**

2 **a** Using integration, show that the work done in stretching a light elastic string of natural length l and modulus of elasticity λ, from length l to length $(l + x)$ is $\dfrac{\lambda x^2}{2l}$

b The same string is stretched from a length $(l + a)$ to a length $(l + b)$ where $b \geqslant a$. Show that the work done is the product of the mean tension and the distance moved.

← **Mechanics 3 Section 2.2**

4 CIRCULAR MOTION

Learning objectives

After completing this chapter you should be able to:

● Understand and calculate angular speed of an object moving in a circle

→ pages 85–88

● Understand and calculate angular acceleration of an object moving on a circular path

→ pages 89–94

● Solve problems with objects moving in horizontal circles

→ pages 95–102

● Solve problems with objects moving in vertical circles

→ pages 103–109

● Solve problems when objects do not stay on a circular path

→ pages 110–119

Prior knowledge check

1 A smooth ring is threaded (i.e. passed onto a long thin piece of string or thread) on a light inextensible string. The ends of the string are attached to a horizontal ceiling, and make angles of 30° and 60° with the ceiling respectively. The ring is held in equilibrium by a horizontal force of magnitude 8 N.

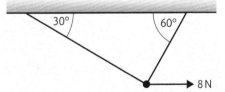

Find
a the tension in the string   b the mass of the ring.

← Mechanics 1 Section 7.2

2 A box of mass 4 kg is projected with speed 10 m s⁻¹ up the line of greatest slope of a rough plane, which is inclined at an angle of 20° to the horizontal. The coefficient of friction between the box and the plane is 0.15. Find
a the distance travelled by the box before it comes to instantaneous rest
b the work done against friction as the box reaches instantaneous rest.

← Mechanics 2 Section 4.1

A car travelling around a bend can be modelled as a particle on a circular path. Police use models such as this to determine likely speeds of cars following accidents.

4.1 Angular speed

When an object is moving in a straight line, the speed, usually measured in m s⁻¹ or km h⁻¹, describes the rate at which distance is changing. For an object moving on a circular path, you can use the same method for measuring speed, but it is often simpler to measure the speed by considering the rate at which the radius is turning.

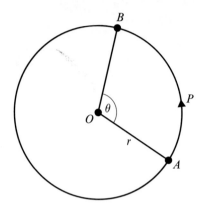

As the particle P moves from point A to point B on the circumference of a circle of radius r m, the radius of the circle turns through an angle θ radians.

The distance moved by P is $r\theta$ m, so if P is moving at v m s⁻¹

we know that $v = \dfrac{\mathrm{d}}{\mathrm{d}t}(r\theta) = r\dfrac{\mathrm{d}\theta}{\mathrm{d}t} = r \times \dot{\theta}$

Notation

$\dot{\theta}$ is the rate at which the radius is turning about O.

It is called the **angular speed of the particle** about O.

The angular speed of a particle is usually denoted by ω, and measured in rad s⁻¹.

- If a particle is moving around a circle of radius r m with linear speed v m s⁻¹ and angular speed ω rad s⁻¹ then $v = r\omega$

Example **1**   **SKILLS** PROBLEM-SOLVING

A particle moves in a circle of radius 4 m with speed 2 m s⁻¹. Calculate the angular speed.

Using $v = r\omega$, $2 = 4\omega$, so $\omega = 0.5$ rad s⁻¹

Example **2**   **SKILLS** PROBLEM-SOLVING

Express an angular speed of 200 **revolutions** per minute in radians per second.

Each complete revolution is 2π radians, so 200 revolutions is 400π radians per minute. Therefore the angular speed is

$\dfrac{400\pi}{60} = 20.9$ rad s⁻¹ (3 s.f.)

Watch out   Sometimes an angular speed is described in terms of the number of revolutions completed in a given time.

Example **3** SKILLS PROBLEM-SOLVING

A particle moves round a circle in 10 seconds at a constant speed of $15\,\text{m}\,\text{s}^{-1}$.
Calculate the angular speed of the particle and the radius of the circle.

The particle rotates through an angle of 2π radians
in 10 seconds, so $\omega = \dfrac{2\pi}{10} = 0.628\,\text{rad}\,\text{s}^{-1}$ (3 s.f.)

Using $v = r\omega$, $r = \dfrac{v}{\omega} = \dfrac{15}{0.628} = 23.9\,\text{m}$ (3 s.f.)

Exercise **4A** SKILLS PROBLEM-SOLVING

1 Express
   a an angular speed of 5 revolutions per minute in $\text{rad}\,\text{s}^{-1}$
   b an angular speed of 120 revolutions per minute in $\text{rad}\,\text{s}^{-1}$
   c an angular speed of $4\,\text{rad}\,\text{s}^{-1}$ in revolutions per minute
   d an angular speed of $3\,\text{rad}\,\text{s}^{-1}$ in revolutions per hour.

2 Find the speed in $\text{m}\,\text{s}^{-1}$ of a particle moving on a circular path of radius $20\,\text{m}$ at
   a $4\,\text{rad}\,\text{s}^{-1}$
   b $40\,\text{rev}\,\text{min}^{-1}$

3 A particle moves on a circular path of radius $25\,\text{cm}$ at a constant speed of $2\,\text{m}\,\text{s}^{-1}$.
   Find the angular speed of the particle
   a in $\text{rad}\,\text{s}^{-1}$
   b in $\text{rev}\,\text{min}^{-1}$

4 Find the speed in $\text{m}\,\text{s}^{-1}$ of a particle moving on a circular path of radius $80\,\text{cm}$ at
   a $2.5\,\text{rad}\,\text{s}^{-1}$
   b $25\,\text{rev}\,\text{min}^{-1}$

5 An athlete is running round a circular track of radius $50\,\text{m}$ at $7\,\text{m}\,\text{s}^{-1}$.
   a How long does it take the athlete to complete one circuit (i.e. complete lap) of the track?
   b Find the angular speed of the athlete in $\text{rad}\,\text{s}^{-1}$.

6 A disc of radius $12\,\text{cm}$ **rotates** at a constant angular speed, completing one revolution every 10 seconds. Find
   a the angular speed of the disc in $\text{rad}\,\text{s}^{-1}$
   b the speed of a particle on the outer rim of the disc in $\text{m}\,\text{s}^{-1}$
   c the speed of a particle at a point $8\,\text{cm}$ from the centre of the disc in $\text{m}\,\text{s}^{-1}$.

7 A cyclist completes two circuits of a circular track in 45 seconds. Calculate

    a his angular speed in rad s^{-1}

    b the radius of the track given that his speed is 40 km h^{-1}.

8 Aalia and Bethany are on a fairground roundabout. Aalia is 3 m from the centre and Bethany is 5 m from the centre. If the roundabout completes 10 revolutions per minute, calculate the speeds with which Aalia and Bethany are moving.

9 A model train completes one circuit of a circular track of radius 1.5 m in 26 seconds. Calculate

    a the angular speed of the train in rad s^{-1}

    b the linear speed of the train in m s^{-1}.

10 A train is moving at 150 km h^{-1} round a circular bend of radius 750 m. Calculate the angular speed of the train in rad s^{-1}.

(P) 11 The hour hand on a clock has radius 10 cm, and the minute hand has radius 15 cm. Calculate

    a the angular speed of the end of each hand

    b the linear speed of the end of each hand.

12 The drum of a washing machine has diameter 50 cm. The drum spins at 1200 rev min^{-1}. Find the linear speed of a point on the drum.

13 A gramophone record rotates at 45 rev min^{-1}. Find

    a the angular speed of the record in rad s^{-1}

    b the distance from the centre of a point moving at 12 cm s^{-1}.

(P) 14 The Earth completes one orbit of the sun in a year. Taking the orbit to be a circle of radius 1.5×10^{11} m, and a year to be 365 days, calculate the speed at which the Earth is moving.

(P) 15 A **bead** moves around a hoop of radius r m with angular velocity 1 rad s^{-1}. The bead moves at a speed greater than 5 m s^{-1}. Find the range of possible values for r.

Challenge

Two separate circular turntables, with different radii, are mounted horizontally on a common vertical axis which acts as the centre of rotation for both. The smaller turntable, of radius 18 cm, is **uppermost** and rotates clockwise (i.e. in the same direction as clock hands). The larger turntable has radius 20 cm and rotates anticlockwise(i.e. in the opposite direction). Both turntables have constant angular velocities, with magnitudes in the same ratio as their radii.

A blue dot is placed at a point on the circumference of the smaller turntable, and a red dot at the same point on the larger one. Starting from the instant that the two dots are at their closest possible distance apart, it is known that 10 seconds later these dots are at their maximum distance apart for the first time. Find the exact angular velocity of the larger turntable.

4.2 Acceleration of an object moving on a horizontal circular path

When an object moves round a horizontal circular path at constant speed, the direction of the motion is changing. If the direction is changing, then, although the speed is constant, the velocity is not constant. If the velocity is changing then the object must have an acceleration.

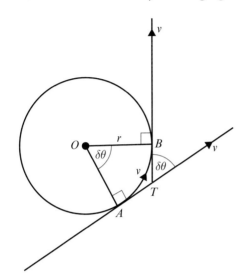

Suppose that the object is moving on a circular path of radius r at constant speed v.

Let the time taken to move from A to B be δt, and the angle AOB be $\delta\theta$.

At A, the velocity is v along the tangent AT.
At B, the velocity is v along the tangent TB.

The velocity at B can be resolved into components

$v \cos \delta\theta$ **parallel** to AT and

$v \sin \delta\theta$ perpendicular to AT.

We know that acceleration $= \dfrac{\text{change in velocity}}{\text{time}}$, so to find the acceleration of the object at the instant when it passes point A, we need to consider what happens to $\dfrac{v \cos \delta\theta - v}{\delta t}$ and $\dfrac{v \sin \delta\theta - 0}{\delta t}$ as $\delta t \to 0$

These will be the components of the acceleration parallel to AT and perpendicular to AT respectively.

For a small angle $\delta\theta$ measured in radians, $\cos \delta\theta \approx 1$ and $\sin \delta\theta \approx \delta\theta$, so the acceleration parallel to AT is zero, and the acceleration perpendicular to AT is $v\dfrac{\delta\theta}{\delta t} = v\omega$

Using $v = r\omega$, $v\omega$ can be written as $r\omega^2$ or $\dfrac{v^2}{r}$

- **An object moving on a circular path with constant linear speed v and constant angular speed ω has acceleration $r\omega^2$ or $\dfrac{v^2}{r}$ towards the centre of the circle.**

Example ④ SKILLS PROBLEM-SOLVING

A particle is moving on a horizontal circular path of radius 20 cm with constant angular speed 2 rad s^{-1}. Calculate the acceleration of the particle.

The radius needs to be measured in metres if the answer is to be in m s^{-2}.

Acceleration
$= 0.2 \times 2^2$
$= 0.8 \text{ m s}^{-2}$ towards the centre of the circle.

Using $a = r\omega^2$

Example 5 **SKILLS** PROBLEM-SOLVING

A particle of mass 150 g moves in a horizontal circle of radius 50 cm at a constant speed of $4\,\mathrm{m\,s^{-1}}$. Find the force towards the centre of the circle that must act on the particle.

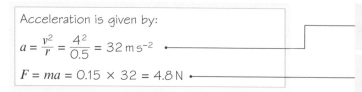

Acceleration is given by:

$a = \dfrac{v^2}{r} = \dfrac{4^2}{0.5} = 32\,\mathrm{m\,s^{-2}}$

$F = ma = 0.15 \times 32 = 4.8\,\mathrm{N}$

Write down the formula for acceleration in terms of speed and radius.

Make sure lengths are in metres and masses are in kg before substituting.

Example 6 **SKILLS** PROBLEM-SOLVING

One end of a light inextensible string of length 20 cm is attached to a particle P of mass 250 g. The other end of the string is attached to a fixed point O on a smooth horizontal table. P moves in a horizontal circle centre O at constant angular speed $3\,\mathrm{rad\,s^{-1}}$. Find the tension in the string.

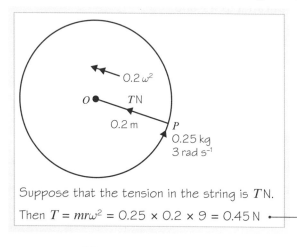

Online Explore circular motion of a particle attached to a light inextensible string using GeoGebra.

The force towards the centre of the circle is due to the tension in the string.

Suppose that the tension in the string is $T\,\mathrm{N}$.

Then $T = mr\omega^2 = 0.25 \times 0.2 \times 9 = 0.45\,\mathrm{N}$

Use $F = ma$ with $a = r\omega^2$

Example 7 **SKILLS** PROBLEM-SOLVING

A smooth wire is formed into a circle of radius 15 cm. A bead of mass 50 g is threaded onto the wire. The wire is horizontal and the bead is made to move along it with a constant speed of $20\,\mathrm{cm\,s^{-1}}$. Find the horizontal component of the force on the bead due to the wire.

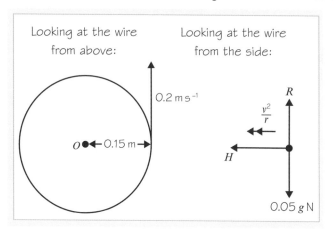

Watch out If a question just says "speed" then it is referring to linear speed.

The forces acting on the bead are weight $0.05g\,\mathrm{N}$, the normal reaction, R, and the horizontal force, H.

The force towards the centre of the circle is due to the horizontal component of the reaction of the wire on the bead.

Let the horizontal component of the force exerted
on the bead by the wire be H.

$H = \dfrac{mv^2}{r} = \dfrac{0.05 \times 0.2^2}{0.15} = 0.013\,N$ (2 s.f.)   •————— Resolve towards the centre of the circle.

Example (8)   **SKILLS** ▸ PROBLEM-SOLVING

A particle P of mass $10\,g$ rests on a rough horizontal disc at a distance $15\,cm$ from the centre.
The disc rotates at constant angular speed of $1.2\,rad\,s^{-1}$, and the particle does not slip.
Calculate the force due to the friction acting on the particle.

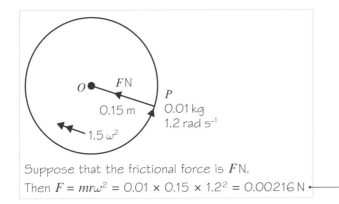

The force towards the centre of the circle
is due to the friction between the particle
and the disc. This is the force that is
providing the angular acceleration of
the particle.

Suppose that the frictional force is $F\,N$.
Then $F = mr\omega^2 = 0.01 \times 0.15 \times 1.2^2 = 0.00216\,N$ •————— Resolve towards the centre of the circle.

Example (9)   **SKILLS** ▸ INTERPRETATION

A car of mass $M\,kg$ is travelling on a flat road round a bend which is an arc of a circle of radius
$140\,m$. The greatest speed at which the car can travel round the bend without slipping is $45\,km\,h^{-1}$.
Find the coefficient of friction between the tyres of the car and the road.

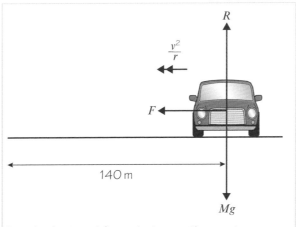

The force towards the centre of the circle is
due to the friction between the tyres of the
car and the road.

Let the frictional force between the car tyres
and the road be F, and the coefficient of friction
be μ. The normal reaction between the car and
the road is R.

Mark the forces on the diagram and
resolve in the direction of the acceleration
and perpendicular to it, i.e. horizontally
and vertically.

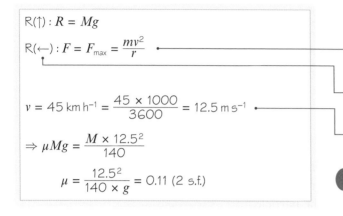

$R(\uparrow): R = Mg$

$R(\leftarrow): F = F_{max} = \dfrac{mv^2}{r}$

$v = 45\,\text{km h}^{-1} = \dfrac{45 \times 1000}{3600} = 12.5\,\text{m s}^{-1}$

$\Rightarrow \mu Mg = \dfrac{M \times 12.5^2}{140}$

$\mu = \dfrac{12.5^2}{140 \times g} = 0.11$ (2 s.f.)

As the car is about to slip at this speed, we know that $F = F_{max} = \mu R$

Resolve towards the centre of the circle.

Convert the speed from km h⁻¹ to m s⁻¹ so that the units are consistent.

Problem-solving

You can cancel M from both sides of the equation. This tells you that the answer is independent of the mass of the car.

Exercise (4B)   **SKILLS**   PROBLEM-SOLVING

Whenever a numerical value of g is required take $g = 9.8\,\text{m s}^{-2}$

1 A particle is moving on a horizontal circular path of radius 16 cm with a constant angular speed of $5\,\text{rad s}^{-1}$. Calculate the acceleration of the particle.

2 A particle is moving on a horizontal circular path of radius 0.3 m at a constant speed of $2.5\,\text{m s}^{-1}$. Calculate the acceleration of the particle.

3 A particle is moving on a horizontal circular path of radius 3 m.
   Given that the acceleration of the particle is $75\,\text{m s}^{-2}$ towards the centre of the circle, find

   a the angular speed of the particle

   b the linear speed of the particle.

4 A particle is moving on a horizontal circular path of diameter 1.2 m.
   Given that the acceleration of the particle is $100\,\text{m s}^{-2}$ towards the centre of the circle, find

   a the angular speed of the particle

   b the linear speed of the particle.

5 A car is travelling round a bend which is an arc of a circle of radius 90 m.
   The speed of the car is $50\,\text{km h}^{-1}$. Calculate its acceleration.

6 A car moving along a horizontal road which follows an arc of a circle of radius 75 m has an acceleration of $6\,\text{m s}^{-2}$ directed towards the centre of the circle.
   Calculate the angular speed of the car.

7 One end of a light inextensible string of length 0.15 m is attached to a particle P of mass 300 g. The other end of the string is attached to a fixed point O on a smooth horizontal table. P moves in a horizontal circle centre O at constant angular speed $4\,\text{rad s}^{-1}$. Find the tension in the string.

8 One end of a light inextensible string of length 25 cm is attached to a particle P of mass 150 g. The other end of the string is attached to a fixed point O on a smooth horizontal table. P moves in a horizontal circle centre O at constant speed $9 \, m \, s^{-1}$. Find the tension in the string.

9 A smooth wire is formed into a circle of radius 0.12 m. A bead of mass 60 g is threaded onto the wire. The wire is horizontal and the bead is made to move along it with a constant speed of $3 \, m \, s^{-1}$. Find
   a the vertical component of the force on the bead due to the wire
   b the horizontal component of the force on the bead due to the wire.

10 A particle P of mass 15 g rests on a rough horizontal disc at a distance 12 cm from the centre. The disc rotates at a constant angular speed of $2 \, rad \, s^{-1}$, and the particle does not slip. Calculate
   a the linear speed of the particle
   b the force due to the friction acting on the particle.

11 A particle P rests on a rough horizontal disc at a distance 20 cm from the centre. When the disc rotates at constant angular speed of $1.2 \, rad \, s^{-1}$, the particle is just about to slip. Calculate the value of the coefficient of friction between the particle and the disc.

12 A particle P of mass 0.3 kg rests on a rough horizontal disc at a distance 0.25 m from the centre of the disc. The coefficient of friction between the particle and the disc is 0.25. Given that P is on the point of slipping, find the angular speed of the disc.

13 A car is travelling round a bend on a flat road which is an arc of a circle of radius 80 m. The greatest speed at which the car can travel round the bend without slipping is $40 \, km \, h^{-1}$. Find the coefficient of friction between the tyres of the car and the road.

14 A car is travelling round a bend on a flat road which is an arc of a circle of radius 60 m. The coefficient of friction between the tyres of the car and the road is $\frac{1}{3}$. Find the greatest angular speed at which the car can travel round the bend without slipping.

(P) 15 A centrifuge (i.e. a machine that spins around to separate what it contains, e.g. liquids from solids) consists of a vertical **hollow** cylinder of radius 20 cm rotating about a vertical axis through its centre at $90 \, rev \, s^{-1}$.
   a Calculate the magnitude of the normal reaction between the cylinder and a particle of mass 5 g on the inner surface of the cylinder.
   b Given that the particle remains at the same height on the cylinder, calculate the least possible coefficient of friction between the particle and the cylinder.

(E/P) 16 A fairground ride consists of a vertical hollow cylinder of diameter 5 m which rotates about a vertical axis through its centre. When the ride is rotating at $W \, rad \, s^{-1}$ the floor of the cylinder opens. Without slipping, the people on the ride remain, in contact with the inner surface of the cylinder.
   a Given that the coefficient of friction between a person and the inner surface of the cylinder is $\frac{2}{3}$, find the minimum value for W.   **(5 marks)**
   b State, with a reason, whether this would be a safe speed at which to operate the ride.   **(1 mark)**

(E) **17** Two particles P and Q, both of mass 80 g, are attached to the ends of a light inextensible string of length 30 cm. Particle P is on a smooth horizontal table, the string passes through a small smooth hole in the centre of the table, and particle Q hangs freely below the table at the other end of the string. P is moving on a circular path about the centre of the table at constant linear speed. Find the linear speed at which P must move if Q is in equilibrium 10 cm below the table.

(4 marks)

(E) **18** A car travels travels around a bend on a flat road. The car is modelled as a particle travelling at a constant speed of v m s^{-1} along a path which is an arc of a circle of radius R m.
Given that the car does not slip,

   a find the minimum value for the coefficient of friction between the car and the road, giving your answer in terms of R and g.

(4 marks)

   b Describe one weakness of the model.

(1 mark)

(P) **19** One end of a light extensible string of natural length 0.3 m and modulus of elasticity 10 N is attached to a particle P of mass 250 g. The other end of the string is attached to a fixed point O on a smooth horizontal table. P moves in a horizontal circle centre O at constant angular speed 3 rad s^{-1}. Find the radius of the circle.

(E/P) **20** A particle P of mass 4 kg rests on a rough horizontal disc, centre O, which is rotating at ω rad s^{-1}. The coefficient of friction between the particle and the disc is 0.3. The particle is attached to O by means of a light elastic string of natural length 1.5 m and modulus of elasticity 12 N. The distance OP is 2 m. Given that the particle does not slide across the surface of the disc, find the maximum possible value of ω.   **(7 marks)**

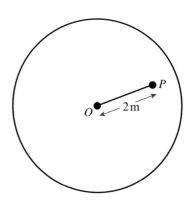

Challenge

A particle is moving in the horizontal x-y plane. Its x- and y-coordinates at time t seconds are given by the parametric equations

$$x = pt \quad y = qt^2 \quad t \geqslant 0$$

where t is the time in seconds, and p and q are positive constants.

a Sketch the path of P and write its equation in the form $y = f(x)$

b Find the acceleration of the particle and its speed, v m s^{-1}, at the origin.

c Find the equation of the lower half of a circle with centre $(0, R)$ and radius R, giving your answer in the form $y = g(x)$

d By comparing second derivatives, find, in terms of p and q, the value of R for which this circle most closely matches the path of P at the origin.

A second particle Q moves around this circle with linear speed v m s^{-1}.

e Find the acceleration of Q.

f Comment on your answer.

4.3 Three-dimensional problems with objects moving in horizontal circles

In this section you will find out how the method of resolving forces can be used to solve a problem about an object moving in a horizontal circle.

Example 10   SKILLS   INTERPRETATION

A particle of mass 2 kg is attached to one end of a light inextensible string of length 50 cm. The other end of the string is attached to a fixed point A. The particle moves with constant angular speed in a horizontal circle of radius 40 cm. The centre of the circle is vertically below A. Calculate the tension in the string and the angular speed of the particle.

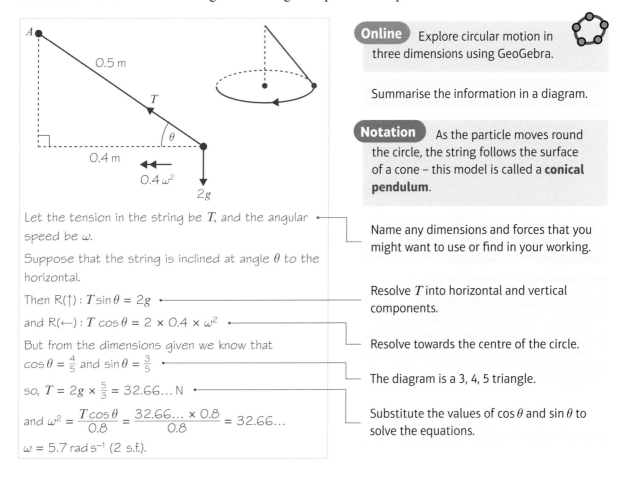

Online Explore circular motion in three dimensions using GeoGebra.

Summarise the information in a diagram.

Notation As the particle moves round the circle, the string follows the surface of a cone – this model is called a **conical pendulum**.

Let the tension in the string be T, and the angular speed be ω.

Suppose that the string is inclined at angle θ to the horizontal.

Then $R(\uparrow): T\sin\theta = 2g$

and $R(\leftarrow): T\cos\theta = 2 \times 0.4 \times \omega^2$

But from the dimensions given we know that $\cos\theta = \frac{4}{5}$ and $\sin\theta = \frac{3}{5}$

so, $T = 2g \times \frac{5}{3} = 32.66... \, \text{N}$

and $\omega^2 = \dfrac{T\cos\theta}{0.8} = \dfrac{32.66... \times 0.8}{0.8} = 32.66...$

$\omega = 5.7 \, \text{rad s}^{-1}$ (2 s.f.).

Name any dimensions and forces that you might want to use or find in your working.

Resolve T into horizontal and vertical components.

Resolve towards the centre of the circle.

The diagram is a 3, 4, 5 triangle.

Substitute the values of $\cos\theta$ and $\sin\theta$ to solve the equations.

Example 11   SKILLS   INTERPRETATION

A particle of mass m is attached to one end of a light inextensible string of length l. The other end of the string is attached to a fixed point A. The particle moves with constant angular speed in a horizontal circle. The string is taut and the angle between the string and the vertical is θ. The centre of the circle is vertically below A. Find the angular speed of the particle.

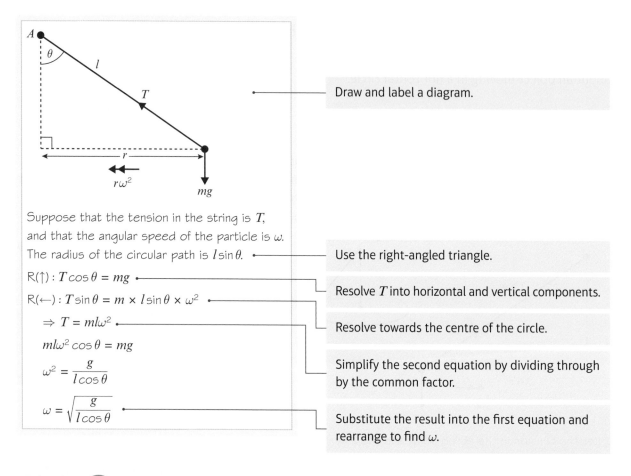

Draw and label a diagram.

Suppose that the tension in the string is T, and that the angular speed of the particle is ω. The radius of the circular path is $l\sin\theta$.

Use the right-angled triangle.

$R(\uparrow): T\cos\theta = mg$

Resolve T into horizontal and vertical components.

$R(\leftarrow): T\sin\theta = m \times l\sin\theta \times \omega^2$

$\Rightarrow T = ml\omega^2$

Resolve towards the centre of the circle.

$ml\omega^2\cos\theta = mg$

$\omega^2 = \dfrac{g}{l\cos\theta}$

Simplify the second equation by dividing through by the common factor.

$\omega = \sqrt{\dfrac{g}{l\cos\theta}}$

Substitute the result into the first equation and rearrange to find ω.

Example **12**   **SKILLS**   PROBLEM-SOLVING

A car travels round a bend of radius 500 m on a flat road which is **banked** at an angle θ to the horizontal. The car is assumed to be moving at constant speed in a horizontal circle and there is no tendency to slip. If there is no frictional force acting on the car down the slope when it is travelling at 90 km h^{-1}, find the value of θ.

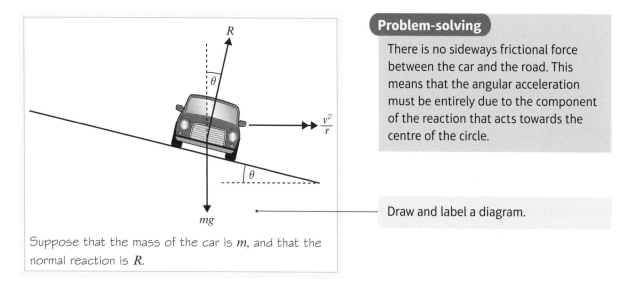

Problem-solving

There is no sideways frictional force between the car and the road. This means that the angular acceleration must be entirely due to the component of the reaction that acts towards the centre of the circle.

Draw and label a diagram.

Suppose that the mass of the car is m, and that the normal reaction is R.

$$90 \, \text{km h}^{-1} = \frac{90 \times 1000}{3600} = 25 \, \text{m s}^{-1}$$

$$R(\uparrow) : R \cos \theta = mg$$

$$R(\leftarrow) : R \sin \theta = \frac{m \times 25^2}{500}$$

$$\Rightarrow \tan \theta = \frac{25^2}{500 \times g} = 0.128\ldots, \; \theta = 7.3° \; (2 \text{ s.f.})$$

Resolve the normal reaction into vertical and horizontal components.

Resolve towards the centre of the circle.

Divide the second equation by the first.

Example 13   SKILLS   CRITICAL THINKING

The diagram shows a particle P of mass m attached by two strings to fixed points A and B, where A is vertically above B. The strings are both taut and P is moving in a horizontal circle with constant angular speed $2\sqrt{3g}$ rad s^{-1}.

Both strings are 0.5 m in length and inclined at 60° to the vertical.

a Calculate the tensions in the two strings.

The strings will break if the tension in them exceeds $8mg$ N.

The angular speed of the particle is increased until the strings break.

b State which string will break first.

c Find the maximum angular speed of the particle before the string breaks.

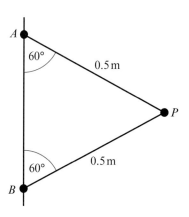

a

Copy the diagram and show all the forces.

The radius of the circular path is

$$0.5 \cos 30° = \frac{\sqrt{3}}{4} \, \text{m}$$

This is an equilateral triangle.

$$R(\uparrow) : T_A \cos 60 = T_B \cos 60 + mg$$

$$\therefore T_A - T_B = 2mg \qquad (1)$$

Resolve both tensions into their horizontal and vertical components.

$$R(\leftarrow) : T_A \cos 30 + T_B \cos 30 = mr\omega^2$$

$$\therefore \frac{\sqrt{3}}{2}(T_A + T_B) = m \times \frac{\sqrt{3}}{4} \times 4 \times 3g$$

Resolve towards the centre of the circle.

$$T_A + T_B = 6mg \qquad (2)$$

$$\Rightarrow T_A = 4mg \, \text{N and} \; T_B = 2mg \, \text{N}$$

Simplify and solve the pair of simultaneous equations (1) and (2).

b The upper string will always have greater tension so will break first.

c Let maximum angular speed be ω_{max}.
At this speed, $T_A = 8\,mg$, so from **(1)**,
$T_B = 6\,mg$

So $\dfrac{\sqrt{3}}{2}(8mg + 6mg) = m\dfrac{\sqrt{3}}{4}\omega_{max}^2$

$28g = \omega_{max}^2$

$\omega_{max} = \sqrt{28g} = 17$ rad s^{-1} (2 s.f.)

> The string will snap if $T_A > 8mg$

> **Problem-solving**
>
> Don't repeat work from part **a** when answering part **c**. All of the working in part **a** up to the point where ω is substituted still applies, so use $T_A - T_B = 2mg$ and $T_A \cos 30° + T_B \cos 30° = mr\omega^2$

Example **14**   **SKILLS**   ADAPTIVE LEARNING

An aircraft of mass 2 tonnes flies at 500 km h^{-1} on a path which follows a horizontal circular arc in order to change course from due north to due east. The aircraft turns in the clockwise direction from due north to due east. It takes 40 seconds to change course, with the aircraft banked at an angle α to the horizontal. Calculate the value of α and the magnitude of the lift force perpendicular to the surface of the aircraft's wings.

> **Problem-solving**
>
> In normal flight, the lift force acts vertically and balances the weight of the aircraft. By banking the aircraft the lift force is now doing two things: the vertical component is balancing the weight, and the horizontal component is the force which causes the acceleration towards the centre of the circular arc that the aircraft is to follow.

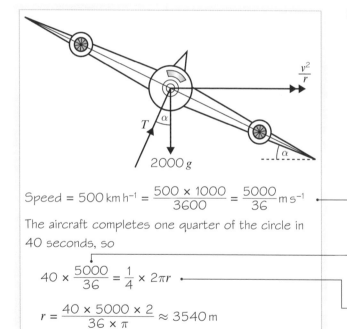

Speed = 500 km h^{-1} = $\dfrac{500 \times 1000}{3600} = \dfrac{5000}{36}$ m s^{-1}

> Convert the speed from km h^{-1} to m s^{-1}.

The aircraft completes one quarter of the circle in 40 seconds, so

$40 \times \dfrac{5000}{36} = \dfrac{1}{4} \times 2\pi r$

> Distance travelled = speed × time

$r = \dfrac{40 \times 5000 \times 2}{36 \times \pi} \approx 3540$ m

> Equate this to one quarter of the circumference of the circle to find the radius of the circle.

$R(\rightarrow): T\sin\alpha = \dfrac{2000 \times \left(\dfrac{5000}{36}\right)^2}{3540} \approx 10\,908$

> Resolve towards the centre of the circle.

$R(\uparrow): T\cos\alpha = 2000g \approx 19\,600$

$\Rightarrow \tan\alpha = \dfrac{10\,908}{19\,600} \approx 0.557, \ \alpha \approx 29°$

> Resolve horizontally and vertically to form two equations in T and α.

and $T \approx \dfrac{19\,600}{\cos\alpha} = 22\,400$ N

> Solve the simultaneous equations.

 **Example** **15**   **SKILLS**   CRITICAL THINKING

In this question use $g = 9.81 \text{ m s}^{-2}$

A hollow right circular cone is fixed with its axis of symmetry vertical and its **vertex** V pointing downwards. A particle, P of mass 25 g moves in a horizontal circle with centre C, and radius 0.27 m, on the rough inner surface of the cone. P remains in contact as it moves with constant angular speed, ω, and does not slip. The angle between VP and the vertical is θ, such that $\tan \theta = 0.45$

The coefficient of friction between the particle and the cone is 0.15

Find the greatest possible value of ω.

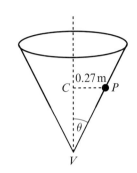

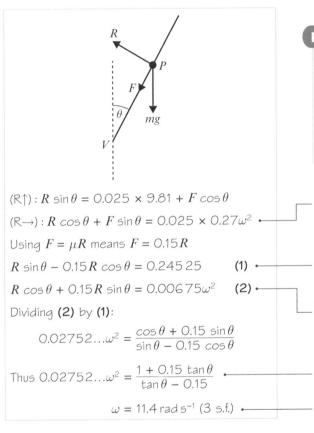

Problem-solving

Begin by drawing a cross-section showing the three forces acting on the particle whilst in motion. You are looking for the **greatest** possible value of ω, so the particle is on the point of slipping **up** the side of the cone. This means that the frictional force acts down towards V.

$(R\uparrow) : R \sin \theta = 0.025 \times 9.81 + F \cos \theta$

$(R\rightarrow) : R \cos \theta + F \sin \theta = 0.025 \times 0.27 \omega^2$

— The angular acceleration is due to the horizontal components of the normal reaction and the frictional force.

Using $F = \mu R$ means $F = 0.15R$

$R \sin \theta - 0.15R \cos \theta = 0.245\,25$ **(1)**

— Resolve vertically and horizontally to form two simultaneous equations.

$R \cos \theta + 0.15R \sin \theta = 0.006\,75\omega^2$ **(2)**

— Substitute to eliminate F.

Dividing **(2)** by **(1)**:

$$0.02752... \omega^2 = \frac{\cos \theta + 0.15 \sin \theta}{\sin \theta - 0.15 \cos \theta}$$

Thus $0.02752... \omega^2 = \dfrac{1 + 0.15 \tan \theta}{\tan \theta - 0.15}$

— Divide the numerator and denominator by $\cos \theta$.

$\omega = 11.4 \text{ rad s}^{-1}$ (3 s.f.) — Use the fact that $\tan \theta = 0.45$

Exercise **4C**   **SKILLS**   INTERPRETATION

Whenever a numerical value of g is required take $g = 9.8 \text{ m s}^{-2}$

1 A particle of mass 1.5 kg is attached to one end of a light inextensible string of length 60 cm. The other end of the string is attached to a fixed point A. The particle moves with constant angular speed in a horizontal circle of radius 36 cm. The centre of the circle is vertically below A. Calculate the tension in the string and the angular speed of the particle.

2 A particle of mass 750 g is attached to one end of a light inextensible string of length 0.7 m. The other end of the string is attached to a fixed point A. The particle moves with constant angular speed in a horizontal circle whose centre is 0.5 m vertically below A. Calculate the tension in the string and the angular speed of the particle.

3 A particle of mass 1.2 kg is attached to one end of a light inextensible string of length 2 m. The other end of the string is attached to a fixed point A. The particle moves in a horizontal circle with constant angular speed. The centre of the circle is vertically below A. The particle takes 2 seconds to complete one revolution. Calculate the tension in the string and the angle between the string and the vertical, to the nearest degree.

4 A conical pendulum consists of a light inextensible string AB of length 1 m, fixed at A and carrying a small ball of mass 6 kg at B. The particle moves in a horizontal circle, with centre vertically below A, at constant angular speed 3.5 rad s^{-1}. Find the tension in the string and the radius of the circle.

(E/P) **5** A conical pendulum consists of a light inextensible string AB of length l, fixed at A and carrying a small ball of mass m at B. The particle moves in a horizontal circle, with centre vertically below A, at constant angular speed ω. Find, in terms of m, l and ω, the tension in the string. **(5 marks)**

(E/P) **6** A conical pendulum consists of a light inextensible string AB fixed at A and carrying a small ball of mass m at B. With the string taut the particle moves in a horizontal circle at constant angular speed ω. The centre of the circle is at distance x vertically below A. Show that $\omega^2 x = g$ **(5 marks)**

(P) **7** A hemispherical bowl of radius r cm is resting in a fixed position with its rim horizontal. A small marble of mass m is moving in a horizontal circle around the smooth inside surface of the bowl. The plane of the circle is 3 cm below the plane of the rim of the bowl. Find the angular speed of the marble.

Problem-solving

The normal reaction of the bowl on the marble will act towards the centre of the sphere.

(P) **8** A hemispherical bowl of radius 15 cm is resting in a fixed position with its rim horizontal. A particle P of mass m is moving at 14 rad s^{-1} in a horizontal circle around the smooth inside surface of the bowl. Find the distance d of the plane of the circle below the plane of the rim of the bowl.

9 A cone is fixed with its base horizontal and its vertex 4 m below the centre of the base. The base has a diameter of 8 m. A particle moves around the smooth inside of the cone a vertical distance 1 m below the base on a horizontal circle. Find the angular and linear speed of the particle.

(E/P) **10** A particle P is moving in a horizonal circle, with centre C
and radius r. P is in contact with the rough inside surface
of a hollow right circular cone. The cone is fixed with its
axis of symmetry vertical and its vertex V pointing
downwards. The radius at the top of the cone is 6 m and
the cone has a perpendicular height of 2 m.

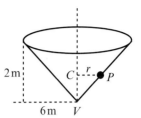

When $r = 0.1$ m, the maximum constant angular speed at which the particle can move,
without slipping from its path, is $14\sqrt{5}$ rad s^{-1}.
Find the maximum angular speed without slipping for $r = 0.3$ m   **(10 marks)**

11 A car travels round a bend of radius 750 m on a road which is banked at an angle θ to the
horizontal. The car is assumed to be moving at constant speed in a horizontal circle and there is
no tendency to slip. If there is no sideways frictional force acting on the car when it is travelling
at 126 km h^{-1}, find the value of θ.

12 A car travels round a bend of radius 300 m on a road which is banked at an angle of 10° to the
horizontal. The car is assumed to be moving at constant speed in a horizontal circle and there is
no tendency to slip. Given that there is no sideways friction acting on the car, find the speed of
the car.

13 A cyclist rides round a circular track of diameter 50 m. The track is banked at 20° to the
horizontal. There is no sideways force due to friction and there is no tendency to slip. By
modelling the cyclist and bicycle as a particle of mass 75 kg, find the speed at which the cyclist
is moving.

(E) **14** A bend in the road is modelled as a horizontal circular arc of radius r. The surface of the
bend is banked at an angle α to the horizontal, and the sideways friction between the tyres and
the road is modelled as being negligible. When a vehicle is driven round the bend there is no
tendency to slip.

    a Show that according to this model, the speed of the vehicle is given by $\sqrt{rg\tan\alpha}$ **(5 marks)**

    b Suggest, with reasons, which modelling assumption is likely to give rise to the greatest
inaccuracy in this calculation.   **(1 mark)**

15 A girl rides her cycle round a circular track of diameter 60 m. The track is banked at 15° to
the horizontal. The coefficient of friction between the track and the tyres of the cycle is 0.25.
Modelling the girl and her cycle as a particle of mass 60 kg moving in a horizontal circle, find
the minimum speed at which she can travel without slipping.

(E) **16** A van is moving on a horizontal circular bend in the road of radius 75 m. The bend is banked
at $\arctan\frac{1}{3}$ to the horizontal. The maximum speed at which the van can be driven round the
bend without slipping is 90 km h^{-1}. Calculate the coefficient of friction between the road surface
and the tyres of the van.   **(4 marks)**

17 A car moves on a horizontal circular path round a banked bend in a race track. The radius of the path is 100 m. The coefficient of friction between the car tyres and the track is 0.3. The maximum speed at which the car can be driven round the bend without slipping is 144 km h^{-1}. Find the angle at which the track is banked, to the nearest degree.

(E/P) **18** A bend in a race track is banked at 30°. A car will follow a horizontal circular path of radius 70 m round the bend. The coefficient of friction between the car tyres and the track surface is 0.4. Find the maximum and minimum speeds at which the car can be driven round the bend without slipping. **(10 marks)**

(E/P) **19** An aircraft of mass 2 tonnes flies at 400 km h^{-1} on a path which follows a horizontal circular arc in order to change course from a bearing of 060° to a bearing of 015°. It takes 25 seconds to change course, with the aircraft banked at $\alpha°$ to the horizontal.

    a Calculate the two possible values of α, to the nearest degree and the corresponding values of the magnitude of the lift force perpendicular to the surface of the aircraft's wings. **(4 marks)**

    b Without further calculation, state how your answers will change if the aircraft wishes to complete its turn in a shorter time. **(3 marks)**

(E/P) **20** A particle of mass m is attached to one end of a light, inextensible string of length l. The other end of the string is attached to a point vertically above the vertex of a smooth cone. The cone is fixed with its axis vertical, as shown in the diagram. The semi-vertical angle of the cone is θ, and the string makes a constant angle of θ with the horizontal, where $\dfrac{\pi}{4} < \theta < \dfrac{\pi}{2}$

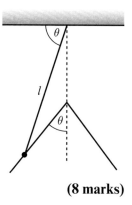

Given that the particle moves in a horizontal circle with angular speed ω, show that the tension in the string is given by $\frac{1}{2}m(\omega^2 l + g \operatorname{cosec} \theta)$ **(8 marks)**

(E/P) **21** A light elastic string AB has natural length 2 m and modulus of elasticity 30 N. The end A is attached to a fixed point. A particle of mass 750 g is attached to the end B. The particle is moving in a horizontal circle below A with the string inclined at 40° to the vertical. Find the angular speed of the particle. **(7 marks)**

4.4 Objects moving in vertical circles

When an object moves in a vertical circle it gains height as it follows its circular path. If it gains height then it must gain gravitational potential energy. Therefore, using the work–energy principle it follows that it must lose kinetic energy, and its speed will not be constant.

You can use **vectors** to understand motion in a vertical circle.

If O is the centre of the circle of radius r and P is the particle, we can set up coordinate axes in the plane of the circle with the x-axis horizontal, and the y-axis vertical.

Let the unit vectors $\mathbf{i}$ and $\mathbf{j}$ be parallel to the x-axis and y-axis respectively.

At time t the angle between the radius OP and the x-axis is θ and the position vector of P is $\mathbf{r}$.

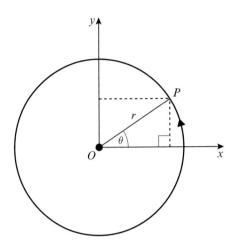

$$\mathbf{r} = (r\cos\theta)\,\mathbf{i} + (r\sin\theta)\,\mathbf{j}$$

By differentiating this with respect to time we obtain the velocity vector

$$\mathbf{v} = \frac{\mathrm{d}}{\mathrm{d}t}(\mathbf{r}) = (-r\sin\theta)\dot\theta\,\mathbf{i} + (r\cos\theta)\dot\theta\,\mathbf{j} = r\dot\theta(-\sin\theta\,\mathbf{i} + \cos\theta\,\mathbf{j})$$

Looking at the directions of $\mathbf{r}$ and $\mathbf{v}$, we find that the lines representing them have gradients $\dfrac{r\sin\theta}{r\cos\theta}$ and $\left(-\dfrac{r\cos\theta}{r\sin\theta}\right)$ respectively.

But $\dfrac{r\sin\theta}{r\cos\theta} \times \left(-\dfrac{r\cos\theta}{r\sin\theta}\right) = -1$, so these two vectors are perpendicular. Alternatively, using the scalar product we see that the vectors are perpendicular since $(\cos\theta\,\mathbf{i} + \sin\theta\,\mathbf{j}) \cdot (-\sin\theta\,\mathbf{i} + \cos\theta\,\mathbf{j}) = 0$

This means that the acceleration has two components, one of magnitude $r\dot\theta^2$ directed towards the centre of the circle, and one of magnitude $r\ddot\theta$ directed along the tangent to the circle.

Using $\dot\theta = \omega$ gives

- For motion in a vertical circle of radius r, the components of the acceleration are $r\omega^2$ or $\dfrac{v^2}{r}$ towards the centre of the circle and $r\ddot\theta = \dot v$ along the tangent.

The force directed towards the centre of the circle is perpendicular to the direction of motion of the particle, so it does no work. If the only other force acting on the particle is gravity, then it follows (using the work–energy principle) that the sum of the kinetic energy and the potential energy of the particle will be constant. You will use this fact to solve problems about motion in a vertical circle.

Links The work-energy principle states that the change in the total energy of a particle is equal to the work done on the particle. This means that where the only force acting on a particle is gravity, the sum of its kinetic and gravitational potential energies remains constant.

← Mechanics 2 Section 4.3

Example **16**   **SKILLS**   PROBLEM-SOLVING

A particle of mass 0.4 kg is attached to one end A of a light rod AB of length 0.3 m. The rod is free to rotate in a vertical plane about B. The particle is held at rest with AB horizontal. The particle is released. Calculate

a the speed of the particle as it passes through the lowest point of the path

b the tension in the rod at this point.

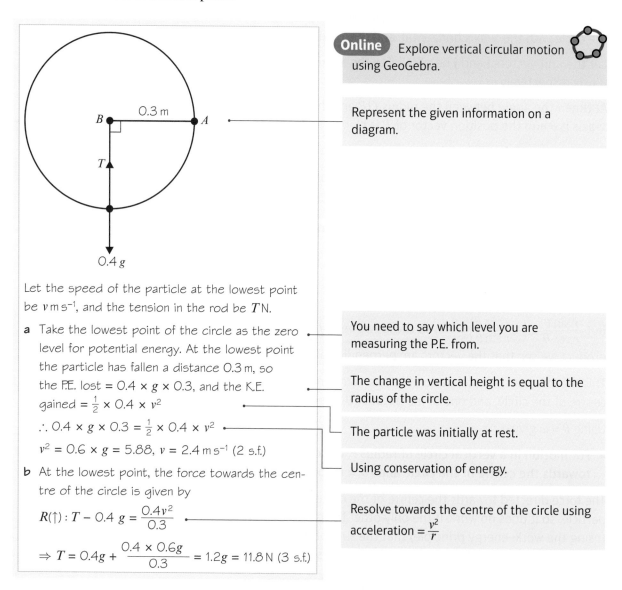

Online   Explore vertical circular motion using GeoGebra.

Represent the given information on a diagram.

Let the speed of the particle at the lowest point be $v\,\text{m s}^{-1}$, and the tension in the rod be $T\,\text{N}$.

a Take the lowest point of the circle as the zero level for potential energy. At the lowest point the particle has fallen a distance 0.3 m, so the P.E. lost = $0.4 \times g \times 0.3$, and the K.E. gained = $\frac{1}{2} \times 0.4 \times v^2$

∴ $0.4 \times g \times 0.3 = \frac{1}{2} \times 0.4 \times v^2$

$v^2 = 0.6 \times g = 5.88$, $v = 2.4\,\text{m s}^{-1}$ (2 s.f.)

b At the lowest point, the force towards the centre of the circle is given by

$R(\uparrow): T - 0.4\,g = \dfrac{0.4v^2}{0.3}$

$\Rightarrow T = 0.4g + \dfrac{0.4 \times 0.6g}{0.3} = 1.2g = 11.8\,\text{N}$ (3 s.f.)

You need to say which level you are measuring the P.E. from.

The change in vertical height is equal to the radius of the circle.

The particle was initially at rest.

Using conservation of energy.

Resolve towards the centre of the circle using acceleration = $\dfrac{v^2}{r}$

Questions about motion in a vertical circle will often ask you to consider whether or not an object will perform complete circles. The next two examples illustrate the importance of considering how the circular motion occurs.

Example **17**   **SKILLS**   PROBLEM-SOLVING

A particle of mass $0.4\,\text{kg}$ is attached to one end A of a light rod AB of length $0.3\,\text{m}$. The rod is free to rotate in a vertical plane about B. The rod is hanging vertically with A below B when the particle is set in motion with a horizontal speed of $u\,\text{m s}^{-1}$. Find

a an expression for the speed of the particle when the rod is at an angle θ to the downward vertical through B

b the minimum value of u for which the particle will perform a complete circle.

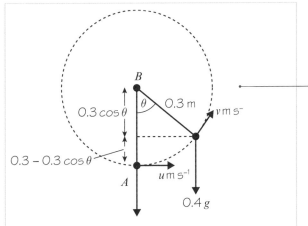

Represent the given information on a diagram.

a Take the lowest point of the circle as the zero level for potential energy.

You need to say which level you are measuring the P.E. from.

At the lowest level the particle has

$$\text{K.E.} = \frac{1}{2} \times 0.4 \times u^2 = 0.2u^2$$

$$\text{P.E.} = 0$$

When the rod is at angle θ to the vertical the particle has

$$\text{K.E.} = \frac{1}{2} \times 0.4 \times v^2 = 0.2v^2$$

$$\text{P.E.} = 0.4 \times g \times 0.3(1 - \cos\theta)$$

$$\therefore 0.2u^2 = 0.2v^2 + 0.12g(1 - \cos\theta)$$

Conservation of energy means that the total energy at each point will be equal.

$$v = \sqrt{u^2 - 0.6g(1 - \cos\theta)}$$

b If the particle is to reach to top of the circle then we require v 0 when $\theta = 180°$

$\cos 180° = -1$

$$\Rightarrow u^2 - 0.6g(1 - \cos 180°) > 0$$

$$u^2 > 0.6g \times 2$$

$$u > \sqrt{1.2g}$$

Hence the minimum value of u is $\sqrt{1.2g}$

Problem-solving

Note that if $u = \sqrt{1.2g}$ then the speed of the particle at the top of the circle would be zero. In this case the rod would be in thrust, with the force in the rod balancing the weight of the particle.

Example 18 **SKILLS** PROBLEM-SOLVING

A particle A of mass $0.4\,\mathrm{kg}$ is attached to one end of a light inextensible string of length $0.3\,\mathrm{m}$. The other end of the string is attached to a fixed point B. The particle is hanging in equilibrium when it is set in motion with a horizontal speed of $u\,\mathrm{m\,s^{-1}}$. Find

a an expression for the tension in the string, in terms of u, when it is at an angle θ to the downward vertical through B

b the minimum value of u for which the particle will perform a complete circle.

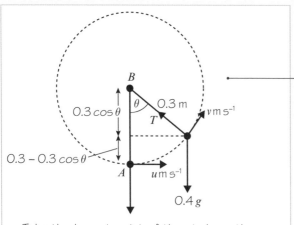

Represent the given information on a diagram.

a Take the lowest point of the circle as the zero level for potential energy.

You need to say which level you are measuring the P.E. from.

At the lowest level the particle has

$$\text{K.E.} = \frac{1}{2} \times 0.4 \times u^2 = 0.2u^2$$
$$\text{P.E.} = 0$$

When the string is at angle θ to the vertical the particle has

$$\text{K.E.} = \frac{1}{2} \times 0.4 \times v^2$$
$$\text{P.E.} = 0.4 \times g \times 0.3(1 - \cos\theta)$$
$$\therefore 0.2u^2 = 0.2v^2 + 0.12g(1 - \cos\theta)$$

Conservation of energy means that the total energy at each point will be equal.

Resolving towards the centre of the circle:

$$R(\nwarrow): T - 0.4g\cos\theta = \frac{mv^2}{r} = \frac{0.4v^2}{0.3}$$

Use $a = \dfrac{v^2}{r}$

$$T = 0.4g\cos\theta + \frac{4}{3}(u^2 - 0.6g + 0.6g\cos\theta)$$

Express v^2 in terms of u^2

$$= 1.2g\cos\theta + \frac{4u^2}{3} - 0.8g$$

b If the particle is to reach to top of the circle then we require $T > 0$ when $\theta = 180°$

$$\Rightarrow -1.2g + \frac{4u^2}{3} - 0.8g > 0$$
$$\frac{4u^2}{3} > 2g$$
$$u^2 > \frac{6g}{4}$$
$$u > \sqrt{\frac{3g}{2}}$$

Hence the minimum value of u is $\dfrac{3g}{2}$

Problem-solving

In the previous example the rod could be in thrust, and could support the particle. In this example the string must remain taut for the particle to perform a complete circle. The condition for the string to remain taut is that the tension on the string remains positive.

Examples 17 and 18 above illustrate the difference between particles attached to strings and rods. You can use these conditions to determine whether particles moving in a vertical circle perform complete circles.

- A particle attached to the end of a light rod will perform complete vertical circles if it has speed > 0 at the top of the circle.

- A small bead threaded on to a smooth circular wire will perform complete vertical circles if it has speed > 0 at the top of the circle.

- A particle attached to a light inextensible string will perform complete vertical circles if the tension in the string > 0 at the top of the circle. This means that the speed of the particle when it reaches the top of the circle must be large enough to keep the string taut at the top of the circle.

 **Exercise** **4D**   **SKILLS**   PROBLEM-SOLVING

Whenever a numerical value of g is required take $g = 9.8\,\mathrm{m\,s^{-2}}$

1 A particle of mass 0.6 kg is attached to end A of a light rod AB of length 0.5 m. The rod is free to rotate in a vertical plane about B. The particle is held at rest with AB horizontal. The particle is released. Calculate
   a the speed of the particle as it passes through the lowest point of the path
   b the tension in the rod at this point.

2 A particle of mass 0.4 kg is attached to end A of a light rod AB of length 0.3 m. The rod is free to rotate in a vertical plane about B. The particle is held at rest with A vertically above B. The rod is slightly displaced so that the particle moves in a vertical circle. Calculate
   a the speed of the particle as it passes through the lowest point of the path
   b the tension in the rod at this point.

3 A particle of mass 0.4 kg is attached to end A of a light rod AB of length 0.3 m. The rod is free to rotate in a vertical plane about B. The particle is held at rest with AB at 60° to the upward vertical. The particle is released. Calculate
   a the speed of the particle as it passes through the lowest point of the path
   b the tension in the rod at this point.

4 A particle of mass 0.6 kg is attached to end A of a light rod AB of length 0.5 m. The rod is free to rotate in a vertical plane about B. The particle is held at rest with AB at 60° to the upward vertical. The particle is released. Calculate
   a the speed of the particle as it passes through the point where AB is horizontal
   b the tension in the rod at this point.

5 A smooth bead of mass 0.5 kg is threaded onto a circular wire ring of radius 0.7 m that lies in a vertical plane. The bead is at the lowest point on the ring when it is projected horizontally with speed 10 m s⁻¹. Calculate
   a the speed of the bead when it reaches the highest point on the ring
   b the reaction of the ring on the bead at this point.

(P)   **6**   A particle of mass 0.5 kg moves around the interior of a sphere of radius 0.7 m. The particle moves in a circle in the vertical plane containing the centre of the sphere. The line joining the centre of the sphere to the particle makes an angle of θ with the vertical. The particle is resting on the bottom of the sphere when it is projected horizontally with speed u m s^{-1}. Find

    a   an expression for the speed of the particle in terms of u and θ

    b   the restriction on u if the particle is to reach the highest point of the sphere.

(E/P)   **7**   A particle A of mass 1.5 kg is attached to one end of a light inextensible string of length 2 m. The other end of the string is attached to a fixed point B. The particle is hanging in equilibrium when it is set in motion with a horizontal speed of u m s^{-1}. Find

    a   an expression for the tension in the string when it is at an angle θ to the downward vertical through B **(3 marks)**

    b   the minimum value of u for which the particle will perform a complete circle.   **(3 marks)**

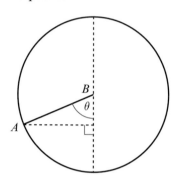

(E/P)   **8**   A small bead of mass 50 g is threaded on a smooth circular wire of radius 75 cm which is fixed in a vertical plane. The bead is at rest at the lowest point of the wire when it is hit with an impulse of I N s horizontally causing it to start to move round the wire. Find the value of I if

    a   the bead just reaches the top of the circle   **(4 marks)**

    b   the bead just reaches the point where the radius from the bead to the centre of the circle makes an angle of $\arctan\frac{3}{4}$ with the upward vertical and then starts to slide back to its original position.   **(3 marks)**

(E/P)   **9**   A particle of mass 50 g is attached to one end of a light inextensible string of length 75 cm. The other end of the string is attached to a fixed point. The particle is hanging at rest when it is hit with an impulse of I N s horizontally causing it to start to move in a vertical circle. Find the value of I if

    a   the particle just reaches the top of the circle   **(4 marks)**

    b   the string goes slack at the instant when the particle reaches the point where the string makes an angle of $\arctan\frac{3}{4}$ with the upward vertical.   **(3 marks)**

    c   Describe the subsequent motion in part **b** qualitatively.   **(1 mark)**

  10   A particle of mass 0.8 kg is attached to end A of a light rod AB of length 2 m. The end B is attached to a fixed point so that the rod is free to rotate in a vertical circle with its centre at B. The rod is held in a horizontal position and then released. Calculate the speed of the particle and the tension in the rod when

    a   the particle is at the lowest point of the circle

    b   the rod makes an angle of $\arctan\frac{3}{4}$ with the downward vertical through B.

  11   A particle of mass 500 g describes complete vertical circles on the end of a light inextensible string of length 1.5 m. Given that the speed of the particle is 8 m s^{-1} at the highest point, find

    a   the speed of the particle when the string is horizontal

    b   the magnitude of the tangential acceleration when the string is horizontal

    c   the tension in the string when the particle is at the lowest point of the circle.

E/P **12** A light rod AB of length 1 m has a particle of mass 4 kg
attached at A. End B is pivoted to a fixed point so that
AB is free to rotate in a vertical plane. When the rod is
vertical with A below B the speed of the particle is $6.5\,\text{m s}^{-1}$.
Find the angle between AB and the vertical at the instant
when the tension in the rod is zero, and calculate the speed
of the particle at that instant. **(7 marks)**

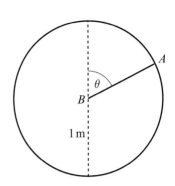

P **13** A particle P of mass m kg is attached to one end of a light rod of length r m which is free to
rotate in a vertical plane about its other end. The particle describes complete vertical circles.
Given that the tension at the lowest point of P's path is three times the tension at the highest
point, find the speed of P at the lowest point on its path.

P **14** A particle P of mass m kg is attached to one end of a light inextensible string of length r m.
The other end of the string is attached to a fixed point O, and P describes complete vertical
circles about O. Given that the speed of the particle at the lowest point is one-and-a-half times
the speed of the particle at the highest point, find
    a the speed of the particle at the highest point
    b the tension in the string when the particle is at the highest point.

E/P **15** A light inelastic string of length r has one end attached to a fixed point O. A particle P of mass
m kg is attached to the other end. P is held with OP horizontal and the string taut. P is then
projected vertically downwards with speed $\sqrt{gr}$
    a Find, in terms of θ, m and g, the tension in the string when OP makes an angle θ with the
        horizontal. **(4 marks)**
    b Given that the string will break when the tension in the string is $2mg$ N, find, to 3 significant
        figures the angle between the string and the horizontal when the string breaks. **(3 marks)**

E/P **16** The diagram shows the cross-section of an industrial roller.
The roller is modelled as a cylinder of radius 4 m. The cylinder
is oriented with its long axis horizontal, and is free to spin
about this axis.
A handle of mass 0.4 kg is attached to the outer surface of
the cylinder at a point S, which is 3.8 m vertically above O.
The cylinder is held in place by this handle, then released
from rest. The handle is modelled as a particle, P.
In the subsequent motion, OP moves in part of a vertical
circle, making an angle θ above the horizontal,

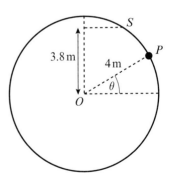

    a show that the linear speed of the handle at any point in its motion is given by
        $\sqrt{7.6g - 8g\sin\theta}$ **(5 marks)**
    b According to the model, state the height of the handle above O at the point where the
        cylinder next comes to rest. **(1 mark)**
    c State, with a reason, how this answer is likely to differ in reality. **(1 mark)**

4.5   Objects not constrained on a circular path

In some models, (for example a bead threaded on a ring or a particle attached to the end of a light rod), the object has to stay on the circular path. If the initial speed is not sufficient for the object to reach the top of the circular path, then it will fall back and oscillate about the lowest point of the path. Other particles may not be constrained (i.e. a forced limit) to stay on a circular path. For example, a particle moving on the convex (i.e. curving out) surface of a sphere.

- If an object is not constrained to stay on its circular path then as soon as the contact force associated with the circular path becomes zero, the object can be treated as a **projectile** moving freely under gravity.

Example **19**   **SKILLS** > **INTERPRETATION**

A particle P of mass m is attached to one end of a light inextensible string of length l. The other end of the string is attached to a fixed point O. The particle is hanging in equilibrium at point A, directly below O, when it is set in motion with a horizontal speed $2\sqrt{gl}$. When OP has turned through an angle θ and the string is still taut, the tension in the string is T. Find

a an expression for T

b the height of P above A at the instant when the string goes slack

c the maximum height above A reached by P before it starts to fall to the ground again.

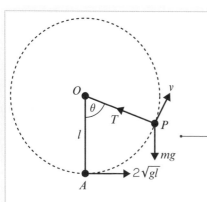

Online Explore motion of a particle not constrained on a circular path using GeoGebra.

Draw and label a diagram.

a When $\angle AOP = \theta$, P has speed v and the tension in the string is T.

Let A be the zero level for P.E.

At A, P has P.E. = 0 and K.E. $= \dfrac{1}{2} \times m \times u^2 = \dfrac{1}{2}m \times 4gl$

When $\angle AOP = \theta$, P has P.E. $= mgl(1 - \cos\theta)$ and — Find the total of P.E. + K.E. at both levels.

$$K.E. = \dfrac{1}{2}mv^2$$

$\therefore 2mgl = mgl(1 - \cos\theta) + \dfrac{1}{2}mv^2$

$v^2 = 2gl(1 + \cos\theta)$ — Energy is conserved.

Resolving parallel to OP:

$$R(\nwarrow) : T - mg\cos\theta = \dfrac{mv^2}{l} = \dfrac{m \times 2gl(1 + \cos\theta)}{l}$$ — Using the equation for circular motion.

$\Rightarrow T = 2mg + 2mg\cos\theta + mg\cos\theta$

$\qquad = 2mg + 3mg\cos\theta$

b When $T = 0$, $\cos \theta = -\dfrac{2}{3}$, so the height of P above A

is $l(1 - \cos \theta) = \dfrac{5l}{3}$.

String slack, so $T = 0$

Substitute for $\cos \theta$

c From the energy equation, we know that when the

string becomes slack $v^2 = 2gl(1 + \cos \theta) = \dfrac{2gl}{3}$

At this point the horizontal component of the velocity

is $v \cos (180 - \theta) = \dfrac{2}{3}\sqrt{\dfrac{2gl}{3}}$

If the additional height before the particle begins to

fall is h, then

$$mgh + \dfrac{1}{2} \times m \times \dfrac{4}{9} \times \dfrac{2gl}{3} = \dfrac{1}{2} \times m \times v^2 = \dfrac{1}{2} \times m \times \dfrac{2gl}{3}$$

,

$$h + \dfrac{4l}{27} = \dfrac{1}{3} \Rightarrow h = \dfrac{5l}{27}$$

$\therefore$ total height above original level $= \dfrac{5l}{27} + \dfrac{5l}{3} = \dfrac{50l}{27}$

Problem-solving

P is now moving freely under gravity. The horizontal component of the velocity will not change.
At the maximum height the vertical component of the velocity is zero.

Conservation of energy.

Watch out The particle is not necessarily directly above A when at its maximum height.

Example **20**   **SKILLS**   PROBLEM-SOLVING

A smooth **hemisphere** with radius 5 m and centre O is resting in a fixed position on a horizontal plane. Its flat face is in contact with the plane. A particle P of mass 4 kg is slightly disturbed from rest at the highest point of the hemisphere.

When OP has turned through an angle θ and the particle is still on the surface of the hemisphere the normal reaction of the sphere on the particle is R. Find

a an expression for R

b the angle between OP and the upward vertical when the particle leaves the surface of the hemisphere

c the distance of the particle from the centre of the hemisphere when it hits the ground.

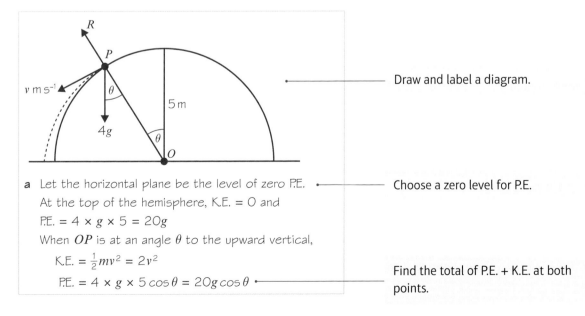

Draw and label a diagram.

a Let the horizontal plane be the level of zero P.E.
At the top of the hemisphere, K.E. $= 0$ and
P.E. $= 4 \times g \times 5 = 20g$
When OP is at an angle θ to the upward vertical,
     K.E. $= \dfrac{1}{2}mv^2 = 2v^2$
     P.E. $= 4 \times g \times 5 \cos \theta = 20g \cos \theta$

Choose a zero level for P.E.

Find the total of P.E. + K.E. at both points.

$\therefore 20g = 2v^2 + 20g\cos\theta$ ●———— **Energy is conserved.**

$v^2 = 10g(1 - \cos\theta)$

Resolving parallel to PO:

$R(\searrow) : 4g\cos\theta - R = \dfrac{mv^2}{r} = \dfrac{4 \times 10g(1 - \cos\theta)}{5}$ ● **Use the equation for circular motion and substitute for v^2.**

$= 8g(1 - \cos\theta)$

so $R = 4g\cos\theta - 8g + 8g\cos\theta = 12g\cos\theta - 8g$

b The particle leaves the hemisphere when $R = 0$

Problem-solving

This is when $\cos\theta = \dfrac{2}{3}$ **The particle leaves the hemisphere when there is no contact force.**

$\theta = \arccos\dfrac{2}{3} = 48.2°$ (3 s.f.)

c When the particle leaves the hemisphere ●———— **The particle is now a projectile with initial velocity $\sqrt{\dfrac{10g}{3}}$ at an angle $\arccos\dfrac{2}{3}$ below the horizontal.**

vertical distance $OP = 5\cos\theta = \dfrac{10}{3}$

horizontal distance $OP = 5\sin\theta = \dfrac{5\sqrt{5}}{3}$

and $v^2 = 10g\left(1 - \dfrac{2}{3}\right) = \dfrac{10g}{3}$

initial vertical speed $= v\sin\theta = \sqrt{\dfrac{10g}{3}} \times \dfrac{\sqrt{5}}{3}$, so

$\dfrac{10}{3} = \sqrt{\dfrac{50g}{27}}t + \dfrac{1}{2}gt^2$

$3\sqrt{3}gt^2 + 2\sqrt{50g}t - 20\sqrt{3} = 0$ ● **Using $s = ut + \dfrac{1}{2}at^2$ and solving the quadratic equation for t.**

$t = 0.4976...$

Horizontal distance travelled in this time

$= v\cos\theta \times t = \sqrt{\dfrac{10g}{3}} \times \dfrac{2}{3} \times 0.4976... = 1.896...$ ●——— **No horizontal acceleration.**

Total distance from $O = \dfrac{5\sqrt{5}}{3} + 1.896... = 5.6\,\text{m}$ (2 s.f.) ● **Add the two horizontal distances.**

Exercise (4E)   **SKILLS**   PROBLEM-SOLVING

Whenever a numerical value of g is required take $g = 9.8\,\text{m}\,\text{s}^{-2}$

(P) **1** A particle P of mass m is attached to one end of a light inextensible string of length l. The other end of the string is attached to a fixed point O. The particle is hanging in equilibrium at a point A, directly below O, when it is set in motion with a horizontal speed $\sqrt{3gl}$. When OP has turned through an angle θ and the string is still taut, the tension in the string is T. Find

  a an expression for T

  b the height of P above A at the instant when the string goes slack

  c the maximum height above A reached by P before it starts to fall to the ground again.

(P) 2 A smooth solid hemisphere with radius 6 m and centre O is resting in a fixed position on a horizontal plane. Its flat face is in contact with the plane. A particle P of mass 3 kg is slightly disturbed from rest at the highest point of the hemisphere.

When OP has turned through an angle θ and the particle is still on the surface of the hemisphere the normal reaction of the sphere on the particle is R. Find

a an expression for R

b the angle, to the nearest degree, between OP and the upward vertical when the particle leaves the surface of the hemisphere

c the distance of the particle from the centre of the hemisphere when it hits the ground.

(P) 3 A smooth solid hemisphere is fixed with its plane face on a horizontal table and its curved surface uppermost. The plane face of the hemisphere has centre O and radius r. The point A is the highest point on the hemisphere. A particle P is placed on the hemisphere at A. It is then given an initial horizontal speed u, where $u^2 = \dfrac{rg}{4}$. When OP makes an angle θ with OA, and while P remains on the hemisphere, the speed of P is v. Find

a an expression for v^2

b the value of $\cos\theta$ when P leaves the hemisphere

c the value of v when P leaves the hemisphere.

After leaving the hemisphere P strikes the table at B, find

d the speed of P at B

e the angle, to the nearest degree, at which P strikes the table.

(P) 4 A smooth sphere with centre O and radius 2 m is fixed to a horizontal surface. A particle P of mass 3 kg is slightly disturbed from rest at the highest point of the sphere and starts to slide down the surface of the sphere. Find

a the angle, to the nearest degree, between OP and the upward vertical at the instant when P leaves the surface of the sphere

b the magnitude and direction, to the nearest degree, of the velocity of the particle as it hits the horizontal surface.

(E/P) 5 A particle of mass m is projected with speed u from the top of the outside of a smooth sphere of radius a. In the subsequent motion the particle slides down the surface of the sphere and leaves the surface of the sphere with speed $\dfrac{\sqrt{3ga}}{2}$. Find

a the vertical distance travelled by the particle before it loses contact with the surface of the sphere   (4 marks)

b u (4 marks)

c the magnitude and direction, to the nearest degree, of the velocity of the particle when it is at the same horizontal level as the centre of the sphere.   (5 marks)

(P) **6** A smooth hemisphere with centre O and radius 50 cm is fixed with its plane face in contact with a horizontal surface. A particle P is released from rest at point A on the sphere, where OA is inclined at $10°$ to the upward vertical. The particle leaves the sphere at point B.

    a Find the angle, to the nearest degree, between OB and the upward vertical.

    b Describe the subsequent motion qualitatively.

(E/P) **7** A smooth laundry chute (i.e. a tube that laundry can be thrown down) is built in two sections, PQ and QR. Each section is in the shape of an arc of a circle. PQ has radius 5 m and subtends an angle of $70°$ at its centre, A. QR has radius 7 m and subtends an angle of $40°$ at its centre, B. The points A, Q and B are in a vertical straight line. The laundry bags are collected in a large bin $\frac{1}{2}$ m below R. To test the chute, a beanbag of mass 2 kg is released from rest at P.

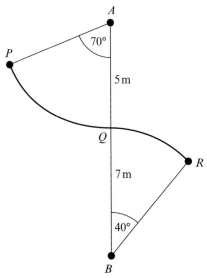

    The beanbag is modelled as a particle and the laundry chute is modelled as being smooth.

    a Calculate the speed with which the beanbag reaches the laundry bin. **(2 marks)**

    b Show that the beanbag loses contact with the chute before it reaches R. **(5 marks)**

    In practice, laundry bags do remain in contact with the chute throughout.

    c State a possible refinement to the model which could account for this discrepancy. **(1 mark)**

(E) **8** Part of a hollow spherical shell, centre O and radius a, is removed to form a smooth bowl with a plane circular rim. The bowl is fixed with the rim uppermost and horizontal. The centre of the circular rim is $\frac{4a}{3}$ vertically above the lowest point of the bowl. A marble is placed inside the bowl and projected horizontally from the lowest point of the bowl with speed u.

    a Find the minimum value of u for which the marble will leave the bowl and not fall back in to it. **(10 marks)**

    In reality the marble is subject to frictional forces from the surface of the bowl and air resistance.

    b State how this will affect your answer to part **a**. **(1 mark)**

Chapter review **4**   **SKILLS** INTERPRETATION

(P) **1** A particle of mass m moves with constant speed u in a horizontal circle of radius $\dfrac{3a}{2}$ on the inside of a fixed smooth hollow sphere of radius $2a$. Show that $9ag = 2\sqrt{7}u^2$

2 A particle P of mass m is attached to one end of a light inextensible string of length $3a$. The other end of the string is attached to a fixed point A which is a vertical distance a above a smooth horizontal table. The particle moves on the table in a circle whose centre O is vertically below A, as shown in the diagram. The string is taut and the speed of P is $2\sqrt{ag}$. Find

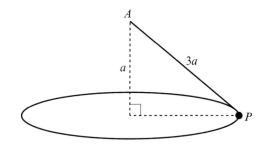

a the tension in the string

b the normal reaction of the table on P.

3 A light inextensible string of length $25l$ has its ends fixed to two points A and B, where A is vertically above B. A small smooth ring of mass m is threaded on the string. The ring is moving with constant speed in a horizontal circle with centre B and radius $12l$, as shown in the diagram. Find

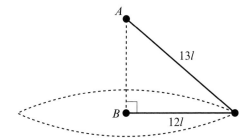

a the tension in the string

b the speed of the ring.

4 A car moves round a bend which is banked at a constant angle of $12°$ to the horizontal. When the car is travelling at a constant speed of $15\,\text{m s}^{-1}$ there is no sideways (i.e. to or from the side) frictional force on the car. The car is modelled as a particle moving in a horizontal circle of radius r metres. Calculate the value of r.

(E/P) **5** A particle P of mass m is attached to the ends of two light inextensible strings AP and BP each of length l. The ends A and B are attached to fixed points, with A vertically above B and $AB = l$, as shown in the diagram. The particle P moves in a horizontal circle with constant angular speed ω. The centre of the circle is the midpoint of AB and both strings remain taut.

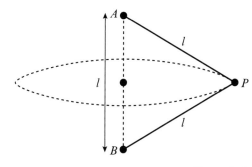

a Show that the tension in AP is $\dfrac{m}{2}(2g + l\omega^2)$ **(3 marks)**

b Find, in terms of m, l, ω and g, an expression for the tension in BP.   **(2 marks)**

c that $\omega^2 > \dfrac{2g}{l}$ **(1 mark)**

P **6** A particle P of mass m is attached to one end of a light string of length l. The other end of the string is attached to a fixed point A. The particle moves in a horizontal circle with constant angular speed ω and with the string inclined at an angle of $45°$ to the vertical, as shown in the diagram.

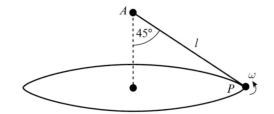

  a Show that the tension in the string is $\sqrt{2}mg$

  b Find ω in terms of g and l.

7 A particle P of mass $0.6\,\text{kg}$ is attached to one end of a light inextensible string of length $1.2\,\text{m}$. The other end of the string is attached to a fixed point A. The particle is moving, with the string taut, in a horizontal circle with centre O vertically below A. The particle is moving with constant angular speed $3\,\text{rad s}^{-1}$. Find

  a the tension in the string

  b the angle, to the nearest degree, between AP and the downward vertical.

E **8** A particle P of mass m moves on the smooth inner surface of a spherical bowl of internal radius r. The particle moves with constant angular speed in a horizontal circle, which is at a depth $\dfrac{r}{4}$ below the centre of the bowl. Find

  a the normal reaction of the bowl on P **(2 marks)**

  b the time it takes P to complete three revolutions of its circular path.   **(4 marks)**

E/P **9** A bend of a race track is modelled as an arc of a horizontal circle of radius $100\,\text{m}$. The track is not banked at the bend. The maximum speed at which a motorcycle can be ridden round the bend without slipping sideways is $21\,\text{m s}^{-1}$. The motorcycle and its rider are modelled as particles.

  a Show that the coefficient of friction between the motorcycle and the track is 0.45.   **(6 marks)**

  The bend is now reconstructed so that the track is banked at an angle α to the horizontal. The maximum speed at which the motorcycle can now be ridden round the bend without slipping sideways is $28\,\text{m s}^{-1}$. The radius of the bend and the coefficient of friction between the motorcycle and the track are unchanged.

  b Find the value of $\tan \alpha$ **(8 marks)**

E/P **10** A light rod rests on the surface of a sphere of radius r, as shown in the diagram. The rod is attached to a point vertically above the centre of the sphere, a distance r from the top of the sphere. A particle, P, of mass m is attached to the rod at the point where the rod meets the sphere. The rod pivots freely such that the particle completes horizontal circles on the smooth outer surface of the sphere with angular speed ω.

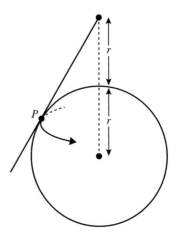

  a Find the tension in the rod above the particle, giving your answer in terms of m, g, ω and r.   **(8 marks)**

Given that the rod remains on the surface of the sphere,

b show that the time taken for the particle to make one complete revolution is at least $\pi\sqrt{\dfrac{6r}{g}}$

(3 marks)

c Without further calculation, state how your answer to part **b** would change if the particle was moved
  i up the rod towards the pivot
  ii down the rod away from the pivot. **(2 marks)**

(E/P) **11** A rough disc rotates in a horizontal plane with constant angular velocity ω about a fixed vertical axis. A particle P of mass m lies on the disc at a distance $\dfrac{3}{5}a$ from the axis. The coefficient of friction between P and the disc is $\dfrac{3}{7}$. Given that P remains at rest relative to the disc,

a prove that $\omega^2 \leqslant \dfrac{5g}{7a}$ **(7 marks)**

The particle is now connected to the axis by a horizontal light elastic string of natural length $\dfrac{a}{2}$ and modulus of elasticity $\dfrac{5mg}{2}$. The disc again rotates with constant angular velocity ω about the axis and P remains at rest relative to the disc at a distance $\dfrac{3}{5}a$ from the axis.

b Find the range of possible values of ω^2. **(8 marks)**

12 A particle P of mass m is attached to one end of a light inextensible string of length a. The other end of the string is fixed at a point O. The particle is held with the string taut and OP horizontal. It is then projected vertically downwards with speed u, where $u^2 = \dfrac{4}{3}ga$. When OP has turned through an angle θ and the string is still taut, the speed of P is v and the tension in the string is T, as shown in the diagram. Find

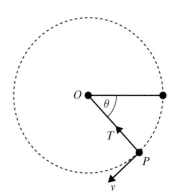

a an expression for v^2 in terms of a, g and θ

b an expression for T in terms of m, g and θ

c the value of θ when the string becomes slack to the nearest degree.

d Explain why P would not complete a vertical circle if the string were replaced by a light rod.

13 A particle P of mass $0.4\,\text{kg}$ is attached to one end of a light inelastic string of length $1\,\text{m}$. The other end of the string is fixed at point O. P is hanging in equilibrium below O when it is projected horizontally with speed $u\,\text{m s}^{-1}$. When OP is horizontal it meets a small smooth peg at Q, where $OQ = 0.8\,\text{m}$. Calculate the minimum value of u if P is to describe a complete circle about Q.

(E/P) **14** A smooth solid hemisphere is fixed with its plane face on a horizontal table and its curved surface uppermost. The plane face of the hemisphere has centre O and radius a. The point A is the highest point on the hemisphere. A particle P is placed on the hemisphere at A.
It is then given an initial horizontal speed u, where $u^2 = \dfrac{ag}{2}$. When OP makes an angle θ with OA, and while P remains on the hemisphere, the speed of P is v.

    a Find an expression for v^2. **(2 marks)**

    b Show that P is still on the hemisphere when $\theta = \arccos 0.9$ **(2 marks)**

    c Find the value of

       i $\cos \theta$ when P leaves the hemisphere

       ii v when P leaves the hemisphere. **(3 marks)**

After leaving the hemisphere P strikes the table at B, find

    d the speed of P at B **(2 marks)**

    e the angle, to the nearest degree, at which P strikes the table. **(3 marks)**

 15 Part of a hollow spherical shell, centre O and radius r, is removed to form a bowl with a plane circular rim. The bowl is fixed with the circular rim uppermost and horizontal. The point C is the lowest point of the bowl. The point B is on the rim of the bowl and OB is at an angle α to the upward vertical as shown in the diagram. Angle α satisfies $\tan \alpha = \dfrac{4}{3}$. A smooth small marble of mass m is placed inside the bowl at C and given an initial horizontal speed u. The direction of motion of the marble lies in the vertical plane COB. The marble stays in contact with the bowl until it reaches B. When the marble reaches B it has speed v.

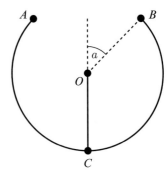

    a Find an expression for v^2. **(4 marks)**

    b If $u^2 = 4gr$ find the normal reaction of the bowl on the marble as the marble reaches B. **(3 marks)**

    c Find the least possible value of u for the marble to reach B. **(3 marks)**

The point A is the other point of the rim of the bowl lying in the vertical plane COB.

    d Find the value of u which will enable the marble to leave the bowl at B and meet it again at A. **(4 marks)**

 16 A particle is at the highest point A on the outer surface of a fixed smooth hemisphere of radius a and centre O. The hemisphere is fixed to a horizontal surface with the plane face in contact with the surface. The particle is projected horizontally from A with speed u, where $u < \sqrt{ag}$. The particle leaves the sphere at the point B, where OB makes an angle θ with the upward vertical, as shown in the diagram.

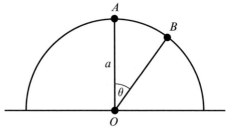

    a Find an expression for $\cos \theta$ in terms of u, g and a. **(3 marks)**

The particle strikes the horizontal surface with speed $\sqrt{\dfrac{5ag}{2}}$

    b Find the value of θ, to the nearest degree. **(4 marks)**

Challenge

The diagram shows the curve with equation $y = f(x)$ $x > 0$, where f is a strictly increasing function.

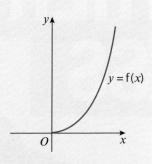

The curve is rotated through 2π radians about the y-axis to form a smooth surface of revolution, which is oriented with the y-axis pointing vertically upwards. A particle is placed on the inside of the surface and completes horizontal circles at a fixed vertical height, with angular speed ω.

a In the case where $f(x) = x^2$ show that ω is independent of the vertical height of the particle, and that $\omega = \sqrt{2g}$

b Conversely (i.e. in an opposite way), show that if ω is independent of the height of the particle, then $f(x)$ must be of the form $px^2 + q$ where p and q are constants.

Problem-solving

f is a strictly increasing function, so $f'(x) > 0$ for all $x > 0$

This means that the particle will be able to complete circles at any position on the inside of the surface, and that the height of the particle will be uniquely determined by its horizontal distance from the y-axis.

Summary of key points

1 If a particle is moving around a circle of radius r m with linear speed v m s^{-1} and angular speed ω rad s^{-1} then $v = r\omega$

2 An object moving on a circular path with constant linear speed v and constant angular speed ω has acceleration $r\omega^2$ or $\frac{v^2}{r}$ towards the centre of the circle.

3 For motion in a vertical circle of radius r, the components of the acceleration are $r\omega^2$ or $\frac{v^2}{r}$ towards the centre of the circle and $r\ddot{\theta} = \dot{v}$ along the tangent.

4 A particle attached to the end of a light rod will perform complete vertical circles if it has speed > 0 at the top of the circle.

5 A small bead threaded on to a smooth circular wire will perform complete vertical circles if it has speed > 0 at the top of the circle.

6 A particle attached to a light inextensible string will perform complete vertical circles if the tension in the string > 0 at the top of the circle. This means that the speed of the particle when it reaches the top of the circle must be large enough to keep the string taut at the top of the circle.

7 If an object is not constrained to stay on its circular path then as soon as the contact force associated with the circular path becomes zero the object can be treated as a projectile moving freely under gravity.

8 For motion in a vertical circle, using the conservation of energy principle will allow you to relate the initial speed with the speed at any point on the circle.

5 FURTHER CENTRES OF MASS

Learning objectives

After completing this chapter you should be able to:

Prior knowledge check

1 Find the area under the curve $y = x^2$ from $x = 1$ to $x = 3$

← Pure 2 Section 8.2

2 The region bounded by the curve with equation
$y = \sqrt{x}\ln x$, the x-axis and the line $x = 4$ is rotated through 360°
about the x-axis. Show that volume of the resulting solid is
$\pi(A \ln 4 - B)$ where A and B are positive rational constants to
be found.

← Pure 4 Section 6.1

3 Given that $y = (1 + 2x)$, evaluate $\dfrac{\displaystyle\int_0^4 xy^2\,dx}{\displaystyle\int_0^4 y^2\,dx}$

← Pure 2 Section 8.1

When a breakdancer balances,
they adjust their body position
so that their centre of mass is
in a favourable position.

5.1 Using calculus to find the centre of mass of a lamina

In the M2 syllabus you found the centres of mass of **laminas** by considering moments and symmetry. For a system of particles m_1, m_2, ... positioned at (x_1, y_1), (x_2, y_2), ... respectively in the plane:

■ $\sum m_i x_i = \bar{x} \sum m_i$ and $\sum m_i y_i = \bar{y} \sum m_i$

You used this result to find the centre of mass of a **composite** lamina by considering the centres of mass of its component parts as particles. You can also use these results in conjunction with integration to find the position of the centre of mass of a uniform lamina.

Suppose you need to find the centre of mass of the uniform lamina bounded by the curve with equation $y = f(x)$, the x-axis and the lines $x = a$ and $x = b$ shown shaded in the diagram below.

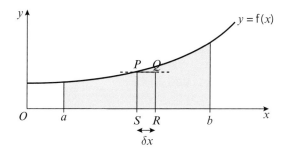

Consider the lamina as made up of small rectangular strips such as $PQRS$, where PQ is parallel to the x-axis. Let P have coordinates (x, y) and let the width of the strip be δx.

> **Notation**   The rectangle $PQRS$ of width δx is sometimes called an **elemental** strip of the lamina.

The height of the strip is y so its area is $y\,\delta x$.

The mass of the strip (m_i) is $\rho y\,\delta x$, where ρ is the mass per unit area of the lamina.

> **Watch out** ρ is the Greek letter 'rho'. In this chapter ρ might represent mass **per unit area**, mass **per unit length** or **density** (mass per **unit volume**). You will be told which quantity it represents in the question.

As the lamina is uniform, as $\delta x \to 0$ the centre of mass of the strip $\to \left(x, \frac{1}{2}y \right)$

Let the coordinates of the centre of mass of the whole lamina be the point $(\bar{x}, \bar{y})$

$$\sum m_i x_i = \bar{x} \sum m_i$$

So $\sum ((\rho y \delta x) x) = \bar{x} \sum \rho y \delta x$

———— The **summation** is taken across all the strips between $x = a$ and $x = b$

$$\bar{x} = \frac{\sum ((\rho y \delta x) x)}{\sum \rho y \delta x}$$

As $\delta x \to 0$ the summations become integrals, giving

$$\bar{x} = \frac{\int_a^b \rho xy \, dx}{\int_a^b \rho y \, dx}$$

$$= \frac{\int_a^b x f(x) \, dx}{\int_a^b f(x) \, dx}$$

Similarly

$$\sum m_i y_i = \bar{y} \sum m_i$$

So $\sum \left((\rho y \delta x) \dfrac{y}{2} \right) = \bar{y} \sum \rho y \delta x$

$$\bar{y} = \frac{\sum \left((\rho y \delta x) \dfrac{y}{2} \right)}{\sum \rho y \delta x}$$

As $\delta x \to 0$ the summations become integrals, giving

$$\bar{y} = \frac{\int_a^b \frac{1}{2} \rho y^2 \, dx}{\int_a^b \rho y \, dx}$$

$$= \frac{\int_a^b \frac{1}{2} (f(x))^2 \, dx}{\int_a^b f(x) \, dx}$$

■ The centre of mass of a uniform lamina may be found using the formulae:

· $\bar{x} = \dfrac{\int_a^b xy \, dx}{\int_a^b y \, dx}$ and $\bar{y} = \dfrac{\int_a^b \frac{1}{2} y^2 \, dx}{\int_a^b y \, dx}$

· $M\bar{x} = \int_a^b \rho xy \, dx$ and $M\bar{y} = \int_a^b \frac{1}{2} \rho y^2 \, dx$, where $M = \int_a^b \rho y \, dx$ is the total mass of the lamina, and ρ is the mass per unit area of the lamina.

Example **1**   SKILLS   PROBLEM-SOLVING

Use calculus to find the position of the centre of mass of a right-angled triangular lamina OPQ with base d and height h, as shown in the diagram.

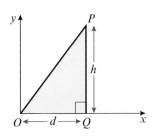

The equation of OP is $y = \dfrac{h}{d}x$

The mass M of the triangular lamina

$$= \rho \times \text{area} = \rho \times \frac{1}{2}dh$$

Using the formula for $\bar{x}$, $M\bar{x} = \displaystyle\int_a^b \rho xy\,dx$

$$M\bar{x} = \int_0^d \rho x \frac{h}{d}x\,dx = \frac{h}{d}\rho \int_0^d x^2\,dx$$

$$= \frac{h}{d}\rho \left[\frac{1}{3}x^3\right]_0^d = \frac{1}{3}\rho hd^2$$

So $\bar{x} = \dfrac{\frac{1}{3}\rho hd^2}{\frac{1}{2}\rho hd} = \dfrac{2}{3}d$

Also $M\bar{y} = \displaystyle\int_a^b \frac{1}{2}\rho y^2\,dx = \int_0^d \frac{1}{2}\rho \left(\frac{h}{d}x\right)^2 dx$

$$= \frac{1}{2}\rho \left(\frac{h}{d}\right)^2\left[\frac{1}{3}x^3\right]_0^d = \frac{1}{6}\rho h^2 d$$

So $\bar{y} = \dfrac{\frac{1}{6}\rho h^2 d}{\frac{1}{2}\rho hd} = \dfrac{1}{3}h$

So the centre of mass is at the point $\left(\dfrac{2}{3}d, \dfrac{1}{3}h\right)$

Find the equation of the line OP by calculating the gradient of OP and using $y = mx + c$, with $c = 0$

Use area of triangle formula and let the mass per unit area be ρ.

Use the formulae for the centre of mass of a lamina to find $\bar{x}$ and $\bar{y}$.

Notice that the centre of mass is at
$$\left(\frac{x_1 + x_2 + x_3}{3}, \frac{y_1 + y_2 + y_3}{3}\right)$$
i.e. $\left(\dfrac{0 + d + d}{3}, \dfrac{0 + 0 + h}{3}\right)$

Example **2**   SKILLS   PROBLEM-SOLVING

Find the coordinates of the centre of mass of the uniform lamina bounded by the curve with equation $y = 4 - x^2$, the x-axis and the y-axis, as shown.

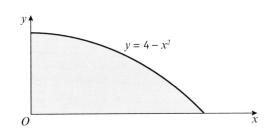

The curve meets the x-axis when $x = 2$ →   Put $y = 0$ and solve $4 - x^2 = 0$ to obtain $x = 2$

$$\bar{x} = \frac{\int_0^b xy\,dx}{\int_0^b y\,dx} = \frac{\int_0^2 x(4 - x^2)\,dx}{\int_0^2 (4 - x^2)\,dx}$$

→   Substitute $y = 4 - x^2$ into the formula for $\bar{x}$.

$$\int_0^2 x(4 - x^2)\,dx = \int_0^2 (4x - x^3)\,dx$$

$$= \left[2x^2 - \frac{1}{4}x^4\right]_0^2 = 8 - 4 = 4$$

$$\int_0^2 (4 - x^2)\,dx = \left[4x - \frac{1}{3}x^3\right]_0^2$$

$$= 8 - \frac{8}{3} = \frac{16}{3}$$

So $\bar{x} = \dfrac{4}{\frac{16}{3}} = \dfrac{3}{4}$ →   Integrate and evaluate $\bar{x}$.

$$\bar{y} = \frac{\int_a^b \frac{1}{2}y^2\,dx}{\int_a^b y\,dx} = \frac{\int_0^2 \frac{1}{2}(4 - x^2)^2\,dx}{\int_0^2 (4 - x^2)\,dx}$$

→   Substitute $y = 4 - x^2$ into the formula for $\bar{y}$.

$$\int_0^2 \frac{1}{2}(4 - x^2)^2\,dx = \frac{1}{2}\int_0^2 (16 - 8x^2 + x^4)\,dx$$

$$= \frac{1}{2}\left[16x - \frac{8}{3}x^3 + \frac{1}{5}x^5\right]_0^2$$

$$= \frac{1}{2}\left(32 - \frac{64}{3} + \frac{32}{5}\right) = \frac{128}{15}$$

→   Integrate and evaluate $\bar{y}$.

So $\bar{y} = \dfrac{\frac{128}{15}}{\frac{16}{3}} = \dfrac{8}{5}$

The coordinates of the centre of mass are
$\left(\dfrac{3}{4}, \dfrac{8}{5}\right)$

Example **3**   **SKILLS**   CRITICAL THINKING

A uniform semicircular lamina has radius r cm. Find the position of its centre of mass.

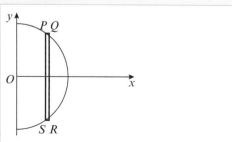

Take the diameter of the lamina as the y-axis, and the midpoint of the diameter as the origin.

Let $PQRS$ be an elemental strip with width δx, where P has coordinates (x, y)

The centre of mass of this strip is at the point $(x, 0)$. — The width of the strip may be ignored as it is small.

The centres of mass of all such strips are on the x-axis and so the centre of mass of the lamina is also on the x-axis. — Use a symmetry argument to explain why the centre of mass lies on the x-axis. This is the axis of symmetry of the lamina.

As point P lies on the circumference of the circle radius r, $x^2 + y^2 = r^2$, and so

$$y = \sqrt{r^2 - x^2}$$

The area of the strip is $2y\,\delta x$ and so its mass is $2\rho y\,\delta x$, where ρ is the mass per unit area of the lamina. — Note that the length of the strip is $2y$, due to the symmetry of the semicircle.

The mass M of the lamina is $\frac{1}{2}\pi r^2 \rho$ and $\bar{x}$ is obtained from: — The area of the semicircle is $\frac{1}{2}\pi r^2$

$$M\bar{x} = \int_0^r 2\rho x \sqrt{(r^2 - x^2)}\,dx$$

$$= \rho \int_0^r 2x(r^2 - x^2)^{\frac{1}{2}}\,dx$$

$$= \rho \left[-\frac{2}{3}\left[r^2 - x^2\right]^{\frac{3}{2}}\right]_0^r$$

$$= \frac{2}{3}\rho r^3$$

The integration may be done by substitution or by inspection using the chain rule in reverse (i.e. in the opposite direction).

$$\text{So } \bar{x} = \frac{\frac{2}{3}\rho r^3}{\frac{1}{2}\rho\pi r^2} = \frac{4r}{3\pi}$$

The centre of mass is on the axis of symmetry at a distance of $\frac{4r}{3\pi}$ from the straight edge diameter.

Watch out You might be asked to prove this result using calculus in your exam. If you are not specifically asked to use calculus or integration, you may quote this result when solving problems.

Example **4**   **SKILLS**   INTERPRETATION

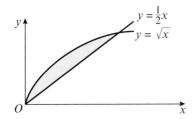

The diagram shows a uniform lamina occupying the shaded region bounded by the curve with equation $y = \sqrt{x}$, and the straight line with equation $y = \frac{1}{2}x$

Find the coordinates of the centre of mass of the lamina.

Consider an elemental strip such as $PQRS$, where P is the point (x, y_1), which lies on the curve $y = \sqrt{x}$, and Q is the point (x, y_2), which lies on the line $y = \frac{1}{2}x$

The area of the strip is $(y_1 - y_2)\,\delta x$ and its mass is $\rho(y_1 - y_2)\,\delta x$ where ρ is the mass per unit area of the lamina.

— δx is the width of the strip.

The centre of mass of the strip lies at the point $\left(x, \frac{1}{2}(y_1 + y_2)\right)$

The line meets the curve when $\sqrt{x} = \frac{1}{2}x$

i.e. when $x = 0$ and $x = 4$

— Square both sides and solve the resulting quadratic equation $x = \frac{1}{4}x^2$

The mass M of the lamina is given by:

$$M = \int_a^b \rho(y_1 - y_2)\,dx = \rho\int_0^4 (\sqrt{x} - \tfrac{1}{2}x)\,dx$$

— Sum the strips and let $\delta x \to 0$, so that the summations become integrals.

$$\text{So } M = \rho\left[\frac{2}{3}x^{\frac{3}{2}} - \frac{1}{4}x^2\right]_0^4 = \rho\left(\frac{16}{3} - \frac{16}{4}\right) = \frac{4}{3}\rho$$

Using $M\bar{x} = \int_a^b \rho x(y_1 - y_2)\,dx = \rho\int_0^4 (x^{\frac{3}{2}} - \tfrac{1}{2}x^2)\,dx$

— Use $M\bar{x} = \sum_{x=a}^{x=b}\rho x(y_1 - y_2)\,\delta x$ and let $\delta x \to 0$, so that the summation becomes an integral.

$$= \rho\left[\frac{2}{5}x^{\frac{5}{2}} - \frac{1}{6}x^3\right]_0^4 = \rho\left(\frac{64}{5} - \frac{64}{6}\right) = \frac{64}{30}\rho$$

So $\bar{x} = \frac{64}{30} \times \frac{3}{4} = \frac{8}{5}$ or 1.6

— Divide $\frac{64}{30}\rho$ by $\frac{4}{3}\rho$, as $m = \frac{4}{3}\rho$

Using $M\bar{y} = \int_a^b \frac{1}{2}\rho(y_1 + y_2)(y_1 - y_2)\,dx$

$$= \frac{1}{2}\rho\int_0^4 \left(\sqrt{x} + \frac{1}{2}x\right)\left(\sqrt{x} - \frac{1}{2}x\right)dx$$

$$= \frac{1}{2}\rho\int_0^4 (x - \tfrac{1}{4}x^2)\,dx$$

— Use $M\bar{y} = \sum_{x=a}^{x=b}\rho\frac{y_1 + y_2}{2}(y_1 - y_2)\,\delta x$ and let $\delta x \to 0$, so that the summation becomes an integral.

$$= \frac{1}{2}\rho\left[\frac{1}{2}x^2 - \frac{1}{12}x^3\right]_0^4 = \frac{1}{2}\rho\left(8 - \frac{64}{12}\right) = \frac{4}{3}\rho$$

So $\bar{y} = \frac{4}{3} \times \frac{3}{4} = 1$

— Divide $\frac{4}{3}\rho$ by $\frac{4}{3}\rho$, as $M = \frac{4}{3}\rho$

The centre of mass is at the point $\left(\frac{8}{5}, 1\right)$

Example (5) **SKILLS** CRITICAL THINKING

Find the centre of mass of a uniform lamina in the form of a sector of a circle, radius r and centre O, which subtends an angle 2α at O.

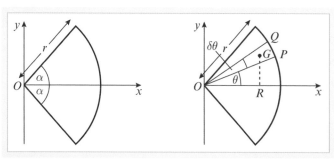

Divide the lamina into **elements** such as OPQ, which is a sector subtending an angle $\delta\theta$ at O.

The area of OPQ is $\frac{1}{2}r^2\,\delta\theta$ and its mass is $\frac{1}{2}r^2\rho\delta\theta$

The sector is approximately a triangle and so its centre of mass, G, is at a distance $\frac{2}{3}r$ from O.

The distance marked OR on the diagram is $\frac{2}{3}r\cos\theta$

The mass M of the whole sector is:

$$\rho \times \frac{1}{2}r^2 2\alpha = \rho r^2\alpha$$

Use $M\bar{x} = \sum_{\theta=-\alpha}^{\theta=\alpha} \frac{1}{2}\rho r^2\,\delta\theta \times \frac{2}{3}r\cos\theta$

As $\delta\theta \to 0$, the summation becomes an integral

and $M\bar{x} = \int_{-\alpha}^{\alpha} \frac{1}{3}\rho r^3 \cos\theta\,d\theta = \frac{1}{3}\rho r^3 \left[\sin\theta\right]_{-\alpha}^{\alpha}$

$$= \frac{1}{3}\rho r^3 2\sin\alpha$$

And so $\bar{x} = \dfrac{\frac{2}{3}\rho r^3 \sin\alpha}{\rho r^2\alpha} = \dfrac{2r\sin\alpha}{3\alpha}$

The distance of the centre of mass from O is $\dfrac{2r\sin\alpha}{3\alpha}$ and it lies on the axis of symmetry.

This is the formula for the area of a sector of a circle.

← **Pure 2 Section 3.2**

The centre of mass of a triangle lies at the intersection of the medians. This is $\frac{2}{3}$ of the way along the line joining each vertex with the midpoint of the opposite side.

Point R is the foot of the perpendicular from G onto the x-axis.

As $\delta\theta$ is small, $\cos\left(\theta + \frac{\delta\theta}{2}\right) \approx \cos\theta$ and this is a reasonable approximation.

This formula is included in the formula booklet, but you should understand and learn how to derive it, as in this example.

Example (6)   **SKILLS** CRITICAL THINKING

Find the centre of mass of a uniform wire in the form of an arc of a circle, radius r and centre O, which subtends an angle 2α at O.

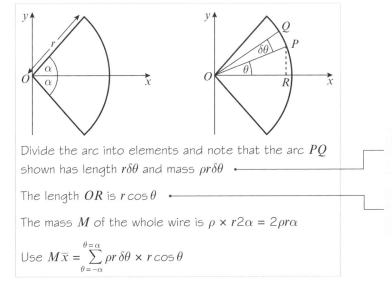

Divide the arc into elements and note that the arc PQ shown has length $r\delta\theta$ and mass $\rho r\delta\theta$

The length OR is $r\cos\theta$

The mass M of the whole wire is $\rho \times r2\alpha = 2\rho r\alpha$

Use $M\bar{x} = \sum_{\theta=-\alpha}^{\theta=\alpha} \rho r\,\delta\theta \times r\cos\theta$

This is the formula for arc length.

← **Pure 2 Section 3.2**

R is the foot of the perpendicular from P to the x-axis.

As $\delta\theta \to 0$ the summation becomes an integral and

$$M\bar{x} = \int_{-\alpha}^{\alpha} \rho r^2 \cos\theta \, d\theta = \rho r^2 \, [\sin\theta]_{-\alpha}^{\alpha} = \rho r^2 \, 2\sin\alpha$$

And so $\bar{x} = \dfrac{2\rho r^2 \sin\alpha}{2\rho r \alpha} = \dfrac{r\sin\alpha}{\alpha}$

The centre of mass lies on the axis of symmetry, and is at a distance $\dfrac{r\sin\alpha}{\alpha}$ from O.

This result is given in the formulae booklet, and you may quote it without proof.

Exercise   SKILLS PROBLEM-SOLVING

1 Find, by integration, the centre of mass of the uniform triangular lamina enclosed by the lines $y = 6 - 3x$, $x = 0$ and $y = 0$

2 Use integration to find the centre of mass of the uniform lamina occupying the **finite** region bounded by the curve with equation $y = 3x^2$, the x-axis and the line $x = 2$

3 Use integration to find the centre of mass of the uniform lamina occupying the finite region bounded by the curve with equation $y = \sqrt{x}$, the x-axis and the line $x = 4$

4 Use integration to find the centre of mass of the uniform lamina occupying the finite region bounded by the curve with equation $y = x^3 + 1$, the x-axis and the line $x = 1$

5 Use integration to find the centre of mass of the uniform lamina occupying the finite region bounded by the curve with equation $y^2 = 4ax$, and the line $x = a$, where a is a positive constant

6 Find the centre of mass of the uniform lamina occupying the finite region bounded by the curve with equation $y = \sin x$, $0 \leqslant x \leqslant \pi$ and the line $y = 0$

7 Find the centre of mass of the uniform lamina occupying the finite region bounded by the curve with equation $y = \dfrac{1}{1+x}$, $0 < x < 1$ and the lines $x = 0$, $x = 1$ and $y = 0$

E/P **8** Find, by integration, the centre of mass of a uniform lamina in the shape of a quadrant of a circle of radius r as shown.

(6 marks)

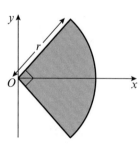

E/P **9** The diagram shows a uniform lamina bounded by the curve $y = x^3$ and the line with equation $y = 4x$, where $x > 0$
Find the coordinates of the centre of mass of the lamina.

(10 marks)

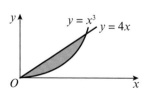

E/P 10 The diagram shows a uniform lamina occupying the finite region bounded by the x-axis, the curve $y = \sqrt{24 - 4x}$ and the line with equation $y = 2x$

Find the coordinates of the centre of mass of the lamina.

(10 marks)

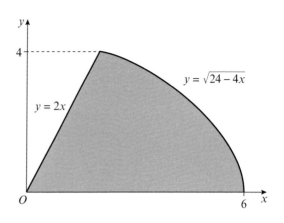

E/P 11 The diagram shows the uniform lamina bounded by the curves with equations $y = \dfrac{x + 5}{(x + 2)(2x + 1)}$ and $y = -\dfrac{x + 5}{(x + 2)(2x + 1)}$, and the lines $x = 0$ and $x = 4$

Find the coordinates of the centre of mass of the lamina.   **(7 marks)**

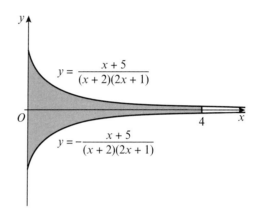

E/P 12 A uniform lamina is bounded by the curve $y = \dfrac{1}{\sqrt{x^2 + 4}}$ the x-axis, and the lines $x = 0$ and $x = 3$, as shown in the diagram.

Find the coordinates of the centre of mass of the lamina.

(10 marks)

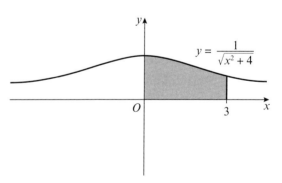

Challenge

The diagram shows the curve C with parametric equations
$x = t^2 \quad y = t \quad t \in \mathbb{R}$

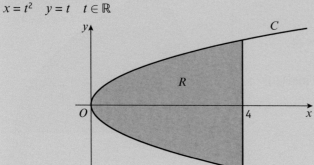

A pendant is modelled as a uniform lamina in the shape of the region R enclosed by the curve and the line $x = 4$

The pendant is suspended in equilibrium from a string attached at a point P on its perimeter, such that no part of the pendant is higher than P.

Find the exact coordinates of the six possible positions of P.

5.2 Centre of mass of a uniform body

You can use symmetry to find the centre of mass of some uniform solids.

- **For a uniform solid body**
  - the weight is evenly distributed through the body
  - the centre of mass will lie on any axis of symmetry
  - the centre of mass will lie on any plane of symmetry

Uniform solid sphere

- **The centre of mass of a uniform solid sphere is at the centre of the sphere.**
  This point is the intersection of the **infinite** number of planes of symmetry and is the only point which lies on all of them.

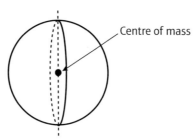

Centre of mass

Uniform solid right circular cylinder

- **The centre of mass of a uniform solid right circular cylinder is at the centre of the cylinder.**
  This point is the intersection of the axis of symmetry and the plane of symmetry which bisects the axis and is parallel to the circular ends.

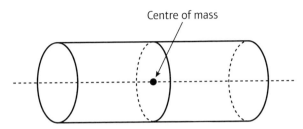

There is another group of solids which are formed by rotating a region through 360° about the x-axis. It is possible to calculate the position of the centres of mass of these solids of revolution by using symmetry and calculus.

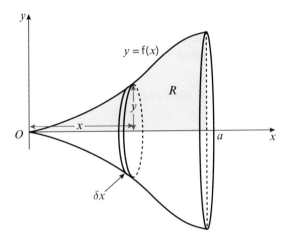

Suppose you need to find the centre of mass of the solid of revolution shown above.

Divide the volume up into a series of very thin circular discs of radius y and thickness δx.

Each disc has a volume of $\pi y^2 \delta x$ and each has a centre of mass that lies on the x-axis at a distance x from O.

So if the distance of the centre of mass of the whole solid of revolution from O is $\bar{x}$, then

$$\bar{x}\sum \rho \pi y^2 \delta x = \sum \rho \pi y^2 x \delta x$$

where ρ is the density, or mass per unit volume

As $\delta x \to 0$, the number of discs becomes infinite and in the limit the sum is replaced by an integral

$$\bar{x} = \frac{\int \rho \pi y^2 x \, dx}{\int \rho \pi y^2 \, dx}$$ ← If the rotation were about the y-axis you would use $\bar{y} = \dfrac{\int \rho \pi x^2 y \, dy}{\int \rho \pi x^2 \, dy}$

which may also be written

$$M\bar{x} = \int \rho \pi y^2 x \, dx$$

where M is the known mass of the solid.

The results from above are summarised below. Note that because the solid is uniform, ρ is constant so it cancels in the formulae for $\bar{x}$ and $\bar{y}$. π is also a constant so it cancels in these formulae.

- For a uniform solid of revolution, where the revolution is about the x-axis, the centre of mass lies on the x-axis and its position on the axis is given by the formulae

$$\bar{x} = \frac{\int y^2 x \, dx}{\int y^2 \, dx} \text{ or } M\bar{x} = \int \rho \pi y^2 x \, dx, \text{ where } M \text{ is the known mass of the solid and } \rho \text{ is its density.}$$

- For a uniform solid of revolution, where the revolution is about the y-axis, the centre of mass lies on the y-axis and its position on the axis is given by the formulae

$$\bar{y} = \frac{\int x^2 y \, dy}{\int x^2 \, dy} \text{ or } M\bar{y} = \int \rho \pi x^2 y \, dy, \text{ where } M \text{ is the known mass of the solid and } \rho \text{ is its density.}$$

This is obtained from the previous result by swapping the values of x and y.

Example ⑦   **SKILLS** ▸ **CRITICAL THINKING**

Find the centre of mass of the uniform solid right circular cone with radius R and height h.

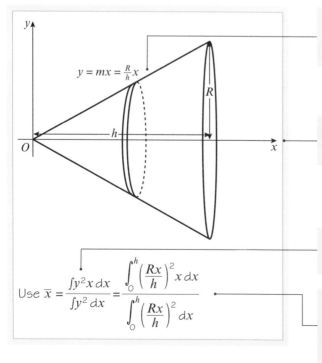

$y = mx = \frac{R}{h}x$

The gradient of the straight line through O is m, where $m = \dfrac{R}{h}$

The centre of mass lies on the axis of symmetry, which is the x-axis in the diagram.

Use $\bar{x} = \dfrac{\int y^2 x \, dx}{\int y^2 \, dx} = \dfrac{\displaystyle\int_0^h \left(\dfrac{Rx}{h}\right)^2 x \, dx}{\displaystyle\int_0^h \left(\dfrac{Rx}{h}\right)^2 \, dx}$

This is a formula you can use if you are finding the centre of mass of a volume of revolution.

The cone is generated by the straight line $y = \dfrac{R}{h}x$, which is rotated through 2π radians about the x-axis.

$$\text{so } \bar{x} = \frac{\dfrac{R^2}{h^2}\displaystyle\int_0^h x^3\,dx}{\dfrac{R^2}{h^2}\displaystyle\int_0^h x^2\,dx}$$

R and h are constant so take $\dfrac{R^2}{h^2}$ outside the integrals and cancel.

$$= \frac{\displaystyle\int_0^h x^3\,dx}{\displaystyle\int_0^h x^2\,dx}$$

$$= \frac{\left[\dfrac{x^4}{4}\right]_0^h}{\left[\dfrac{x^3}{3}\right]_0^h}$$

$$= \frac{\left(\dfrac{h^4}{4}\right)}{\left(\dfrac{h^3}{3}\right)}$$

$$= \frac{3}{4}h$$

Online Explore the centre of mass of a solid of revolution using GeoGebra.

■ The centre of mass of a uniform right circular cone lies on the axis of symmetry and is at a distance $\dfrac{3}{4}h$ from the vertex, or $\dfrac{1}{4}h$ from the circular base.

Example 8 **SKILLS** PROBLEM-SOLVING

Find the centre of mass of the uniform solid hemisphere with radius R.

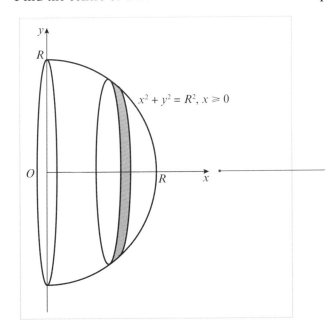

The centre of mass lies on the axis of symmetry, which is the x-axis in the diagram.

Use $\bar{x} = \dfrac{\int y^2 x \, dx}{\int y^2 \, dx}$

$= \dfrac{\displaystyle\int_0^R (R^2 - x^2) x \, dx}{\displaystyle\int_0^R (R^2 - x^2) \, dx}$

$= \dfrac{\int (R^2 x - x^3) \, dx}{\int (R^2 - x^2) \, dx}$

$= \dfrac{\left[R^2 \dfrac{x^2}{2} - \dfrac{x^4}{4} \right]_0^R}{\left[R^2 x - \dfrac{x^3}{3} \right]_0^R}$

$= \dfrac{\left(\dfrac{R^4}{4} \right)}{\left(\dfrac{2R^3}{3} \right)}$

$= \dfrac{3}{8} R$

Divide the sphere up into a series of circular discs. Each disc has mass $\rho \pi y^2 \, \delta x$ and centre of mass at a distance x from O.
So if the distance of the centre of mass of the sphere from O is $\bar{x}$, then, as $\delta x \to 0$

$\bar{x} = \dfrac{\int \rho \pi y^2 x \, dx}{\int \rho \pi y^2 \, dx} = \dfrac{\int y^2 x \, dx}{\int y^2 \, dx}$

The sphere is generated by the circle $y^2 = R^2 - x^2$ which is rotated through 2π radians about the x-axis, so replace y^2 by $R^2 - x^2$

Watch out You might be asked to prove the results for the centre of mass of a cone and sphere using calculus. Otherwise, you can quote these results without proof when solving problems.

■ The centre of mass of a uniform solid hemisphere lies on the axis of symmetry and is at a distance $\dfrac{3}{8} R$ from the plane surface.

Example **9**   **SKILLS**   PROBLEM-SOLVING

Find the centre of mass of the uniform solid of revolution formed by rotating the finite region enclosed by the curve $y = x^2 + 1$, the x-axis and the lines $x = 0$ and $x = 3$ through 360° about the x-axis as shown in the diagram below.

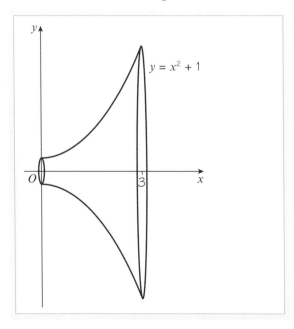

Use $\bar{x} = \dfrac{\int y^2 x \, dx}{\int y^2 \, dx}$

It is a good idea to write out the formula you are using before substituting.

$y = x^2 + 1$ so $y^2 = x^4 + 2x^2 + 1$

$y^2 = (x^2 + 1)^2$

$\bar{x} = \dfrac{\displaystyle\int_0^3 (x^5 + 2x^3 + x)\,dx}{\displaystyle\int_0^3 (x^4 + 2x^2 + 1)\,dx}$

$= \dfrac{\left[\dfrac{1}{6}x^6 + \dfrac{1}{2}x^4 + \dfrac{1}{2}x^2\right]_0^3}{\left[\dfrac{1}{5}x^5 + \dfrac{2}{3}x^3 + x\right]_0^3}$

$= \dfrac{\left[\dfrac{1}{6}(3)^6 + \dfrac{1}{2}(3)^4 + \dfrac{1}{2}(3)^2\right]}{\left[\dfrac{1}{5}(3)^5 + \dfrac{2}{3}(3)^3 + 3\right]}$

$= \dfrac{\dfrac{333}{2}}{\dfrac{348}{5}} = \dfrac{555}{232}$

The centre of mass lies at the point $\left(\dfrac{555}{232}, 0\right)$

Watch out In questions such as this you need to show full algebraic working. This means that you should show the integrated function together with the limits, and the step of substituting the limits. You should not use the numerical integration function on your calculator.

Example (10) **SKILLS** ADAPTIVE LEARNING

Find the centre of mass of the uniform hemispherical shell with radius R.

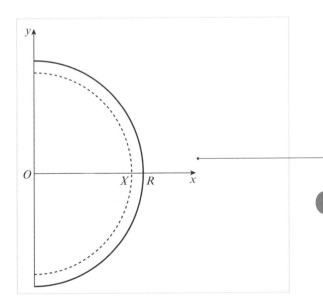

A hemispherical shell is a hollow hemisphere.

Problem-solving

You can obtain a hollow hemisphere by removing a solid concentric hemisphere of radius X from the solid hemisphere of radius R, and then considering what happens as $X \rightarrow R$

Shape	Mass	Centre of mass
Solid hemisphere radius R	$\frac{2}{3}\rho\pi R^3$	$\left(\frac{3}{8}R,\ 0\right)$
Solid hemisphere radius X	$\frac{2}{3}\rho\pi X^3$	$\left(\frac{3}{8}X,\ 0\right)$
Hollow shell	$\frac{2}{3}\rho\pi(R^3 - X^3)$	$(\overline{x},\ 0)$

Taking moments about a horizontal axis through O in the plane face of the hemisphere:

$$\frac{2}{3}\rho\pi R^3 \times \frac{3}{8}R - \frac{2}{3}\rho\pi X^3 \times \frac{3}{8}X = \frac{2}{3}\rho\pi(R^3 - X^3) \times \overline{x}$$

So $\overline{x} = \dfrac{3}{8} \times \dfrac{R^4 - X^4}{R^3 - X^3} = \dfrac{3}{8} \times \dfrac{(R - X)(R + X)(R^2 + X^2)}{(R - X)(R^2 + RX + X^2)}$

$$= \frac{3}{8}\frac{(R + X)(R^2 + X^2)}{(R^2 + RX + X^2)}$$

As $X \to R$ you obtain the result for a hemispherical shell:

$$\overline{x} = \frac{3}{8} \times \frac{(2R)(2R^2)}{(3R^2)}$$

$$= \frac{1}{2}R$$

> From symmetry you can deduce that the centre of mass lies on the x-axis.

> This result is given in the formulae booklet, and you may quote it without proof.

- The centre of mass of a uniform hemispherical shell lies on the axis of symmetry and is at a distance $\frac{1}{2}R$ from the plane surface.

Example 11   **SKILLS**   ADAPTIVE LEARNING

Show that the centre of mass of the uniform hollow right circular cone with radius R and height h is at a distance $\frac{1}{3}h$ from the base along the axis of symmetry.

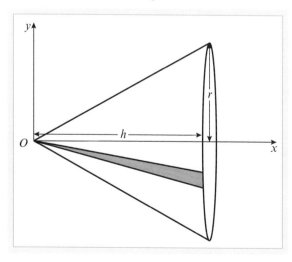

Divide the surface of the cone up into triangular strips with vertices at O and with bases on the circumference of the circular base of the cone. One is shown in the diagram.

Each of these triangles has centre of mass $\frac{2}{3}$ of the distance from O to the base of the cone.

So the centre of mass of the hollow cone is also $\frac{2}{3}$ of the distance from O to the base of the cone, but is on the axis of symmetry.

The distance of the centre of mass from the base is $h - \frac{2}{3}h = \frac{1}{3}h$

Use symmetry to obtain this result.

■ The centre of mass of a conical shell lies on the axis of symmetry and is at a distance $\frac{1}{3}h$ from the base.

Example (12)   **SKILLS** ▷ CREATIVITY

Find the position of the centre of mass of the **frustum** of a right circular uniform cone, of end radii 1 cm and 4 cm, and of height 7 cm.

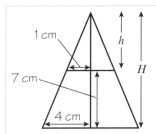

Let the large cone have height H and the small cone have height h.

From similar triangles $\frac{h}{H} = \frac{1}{4}$ or $H = 4h$

But $H = 7 + h$, so $h = \frac{7}{3}$, and $H = \frac{28}{3}$

Shape	Mass	Centre of mass
Large cone	$\frac{1}{3}\rho\pi 4^2 H$ $= \frac{64}{3}\rho\pi h$	$\left(\frac{1}{4}H, 0\right) = \left(\frac{7}{3}, 0\right)$
Small cone	$\frac{1}{3}\rho\pi 1^2 h$ $= \frac{1}{3}\rho\pi h$	$\left(7 + \frac{1}{4}h, 0\right)$ $= \left(\frac{91}{12}, 0\right)$
Frustum	$\frac{63}{3}\rho\pi h$ $= 21\rho\pi h$	$(\bar{x}, 0)$

A frustum is a portion of the cone lying between two parallel planes. It may be considered as a large cone with a small cone removed from the top.

Problem-solving

The masses are in the ratio $64 : 1 : 63$ and you can use these ratios in your moments equation to simplify the working.

So you would have $64 \times \frac{7}{3} - 1 \times \frac{91}{12} = 63\bar{x}$

Taking moments about the base of the frustum:

$$\frac{64}{3}\rho\pi h \times \frac{7}{3} - \frac{1}{3}\rho\pi h \times \frac{91}{12} = 21\rho\pi h\bar{x}$$

So $\bar{x} = \left(\frac{448}{9} - \frac{91}{36}\right) \div 21 = \frac{9}{4}$

The centre of mass is $\frac{9}{4}$ cm above the base on the axis of symmetry.

Exercise SKILLS INTERPRETATION

1 The finite region bounded by the curve $y = x^2 - 4x$ and the x-axis is rotated through 360° about the x-axis to form a solid of revolution. Find the coordinates of its centre of mass.

> **Hint** In questions 1–4 use symmetry to find the coordinates of the centre of mass of the solid.

2 The finite region bounded by the curve $(x - 1)^2 + y^2 = 1$ is rotated through 180° about the x-axis to form a solid of revolution. Find the coordinates of its centre of mass.

3 The finite region bounded by the curve $y = \cos x$, $\frac{\pi}{2} \leqslant x \leqslant \frac{3\pi}{2}$, and the x-axis, is rotated through 360° about the x-axis to form a solid of revolution. Find the coordinates of its centre of mass.

4 The finite region bounded by the curve $y^2 + 6y = x$ and the y-axis is rotated through 360° about the y-axis to form a solid of revolution. Find the coordinates of its centre of mass.

5 Find, by integration, the coordinates of the centre of mass of the solid formed when the finite region bounded by the curve $y = 3x^2$, the line $x = 1$ and the x-axis is rotated through 360° about the x-axis.

> **Hint** In questions 5–10 use integration to find the position of the centre of mass of the solid.

6 Find, by integration, the coordinates of the centre of mass of the solid formed when the finite region bounded by the curve $y = \sqrt{x}$, the line $x = 4$ and the x-axis is rotated through 360° about the x-axis.

7 Find, by integration, the coordinates of the centre of mass of the solid formed when the finite region bounded by the curve $y = 3x^2 + 1$, the lines $x = 0$, $x = 1$ and the x-axis is rotated through 360° about the x-axis.

8 Find, by integration, the coordinates of the centre of mass of the solid formed when the finite region bounded by the curve $y = \frac{3}{x}$, the lines $x = 1$, $x = 3$ and the x-axis is rotated through 360° about the x-axis.

(E) **9** Find, by integration, the coordinates of the centre of mass of the solid formed when the finite region bounded by the curve $y = 2e^x$, the lines $x = 0$, $x = 1$ and the x-axis is rotated through 360° about the x-axis. **(10 marks)**

(E/P) **10** Find, by integration, the coordinates of the centre of mass formed when the region bounded by the curve $xe^y = 3$, the lines $y = 1$, $y = 2$ and the y-axis is rotated through 360° about the y-axis. **(10 marks)**

(E) **11** Find, by integration, the coordinates of the centre of mass formed when the region bounded by the curve $y = \dfrac{2}{1 + x}$, the lines $x = 0$, $x = 4$ and the x-axis is rotated through 360° about the x-axis. **(10 marks)**

(E/P) **12** Find the position of the centre of mass of the frustum of a right circular uniform solid cone, where the frustum has end radii 2 cm and 5 cm, and has height 4 cm. **(8 marks)**

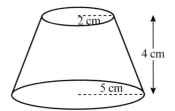

(E/P) **13** The diagram shows a frustum of a right circular uniform cone. The frustum has end diameters of 4 cm and 8 cm, and height 8 cm. A cylindical hole of diameter 2 cm is drilled through the frustum along its axis. Find the distance of the centre of mass of the resulting solid from the larger face of the frustum. **(12 marks)**

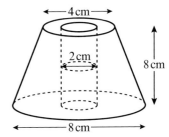

(E/P) **14** A thin uniform hemispherical shell has a circular base of the same material. Find the position of the centre of mass above the base in terms of its radius r. **(6 marks)**

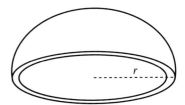

(E/P) **15** A thin uniform hollow cone has a circular base of the same material. Find the position of the centre of mass above the base, given that the radius of the cone is 3 cm and its height is 4 cm. **(6 marks)**

Hint The curved surface area of a cone of slant height l and base radius r has area $\pi r l$.

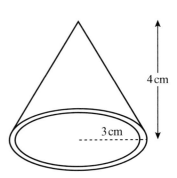

E/P **16** A cap of a sphere is formed by making a plane cut across the
sphere. Show that the centre of mass of a cap of height h cut
from a uniform solid sphere of radius a lies a perpendicular
distance $\dfrac{h(4a - h)}{4(3a - h)}$ from the flat surface of the cap. **(12 marks)**

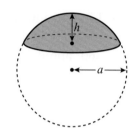

E/P **17** The finite region R, shown shaded on the diagram, is bounded
by the curve with equation $y = 8 - x^3$ and the coordinate axes.

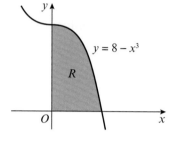

The region is rotated through 2π radians about the y-axis to
form a uniform solid of revolution, S.

a Find, using algebraic integration, the y coordinate of the
centre of mass of S. **(8 marks)**

A solid chocolate egg is formed by attaching a uniform solid
hemisphere to the base of the solid S, as shown in the diagram.
The units of measurement are cm.

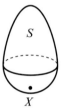

b Find the distance of the centre of mass of the chocolate
egg from its base, X, giving your answer correct to
3 significant figures. **(5 marks)**

E/P **18** A uniform solid is formed from two right circular cones. The cones each
have base radius r and heights x and kx respectively, where k is a constant.
The point O lies at the centre of the common plane face of both cones.

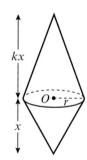

Given that the centre of mass of the solid is a distance $\dfrac{1}{10}x$ above O,
find the value of k. **(5 marks)**

E/P **19** A uniform solid cylinder has height h and radius $2r$.
A hole is drilled in the cylinder in the shape of an
inverted cone of base radius r and height h.
The vertex of the cone lies on the base of the
cylinder, and the axes of the cone and the
cylinder are both vertical. The centre of the
top plane face of the cylinder, O, lies on the
circumference of the base of the cone.

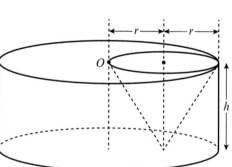

a Find the distance of the centre of mass of the
solid from its top plane face, giving your answer in the form kh where k is a rational constant
to be found. **(7 marks)**

b Fully describe the position of the centre of mass of the solid. **(5 marks)**

20 A game piece is modelled as a solid cone of base radius 5 cm and height 12 cm, sitting on top of a solid cylinder of radius 5 cm and height 3 cm. The cone and cylinder are made of the same uniform material.

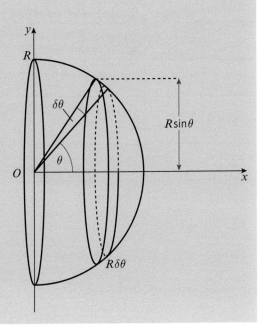

a Find the distance of the centre of mass of the game piece from the base of the cylinder.   **(6 marks)**

In order to save money, the manufacturer decides to make the game piece hollow. The game piece is now modelled as the curved surfaces of a cone and cylinder, and a single circular face on its base. All the surfaces are made from the same uniform material.

b Find the distance of the centre of mass of the hollow game piece from the base of the cylinder.   **(6 marks)**

Challenge

Using calculus, find the centre of mass of the uniform hemispherical shell with radius R.

Hint Divide the shell into small elemental cylindrical rings, centred on the x-axis, with radius $R \sin \theta$, and height $R\,\delta\theta$, where θ is the angle between the radius R and the x-axis.

5.3   **Non-uniform bodies**

You can find the centres of mass of composite bodies made from materials of different densities.

   **SKILLS**   INTERPRETATION

A uniform solid right circular cone of height $2R$ and base radius R is joined at its base to the base of a uniform solid hemisphere. The centres of their bases coincide (i.e. share the same space) at O and their axes are collinear. The radius of the hemisphere is $2R$.

Find the position of the centre of mass of the composite body if the cone has four times the density of the hemisphere.

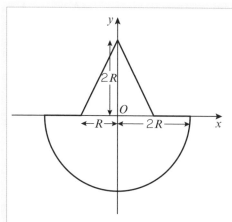

Let ρ be the mass per unit volume for the hemisphere and 4ρ be the mass per unit volume for the cone.

Shape	Mass	Ratio of masses	Distance from O to centre of mass
Cone	$\frac{1}{3}4\rho\pi R^2(2R)$	1	$\frac{1}{4}(2R) = \frac{R}{2}$
Hemisphere	$\frac{2}{3}\rho\pi(2R)^3$	2	$-\frac{3}{8}(2R) = -\frac{3R}{4}$
Composite body	$\frac{1}{3}4\rho\pi R^2(2R) + \frac{2}{3}\rho\pi(2R)^3$	3	$\overline{x}$

The centre of mass lies on the common axis of symmetry.

Take moments about an axis through O:

$1 \times \dfrac{R}{2} - 2 \times \dfrac{3R}{4} = 3 \times \overline{x}$

So $\overline{x} = -\dfrac{1}{3}R$

The centre of mass of the composite body is a distance $-\dfrac{1}{3}R$ below O, on the common axis of symmetry.

Problem-solving

Although each separate solid has uniform density, the composite body does not. Use ρ to represent the density (mass per unit volume) of the hemisphere, so that the cone has density 4ρ.

The masses are in the ratio 1:2:3 and you can use these ratios in your moments equation to simplify the working.

If you take moments about a point inside the body, be careful with your positive and negative signs. The centre of mass of the whole body is **below** O.

Exercise **5C**   **SKILLS**   CRITICAL THINKING

P **1** A uniform solid right circular cone of height 10 cm and base radius 5 cm is joined at its base to the base of a uniform solid hemisphere. The centres of their bases coincide and their axes are collinear. The radius of the hemisphere is also 5 cm. The density of the hemisphere is twice the density of the cone. Find the position of the centre of mass of the composite body.

P **2** A solid is composed of a (i.e. made of) uniform solid right circular cylinder of height 10 cm and base radius 6 cm joined at its top plane face to the base of a uniform cone of the same radius and of height 5 cm. The centres of their **adjoining** circular faces coincide at point O and their axes are collinear. The mass per unit volume of the cylinder is three times that of the cone. Find the position of the centre of mass of the composite body.

(E/P) **3 a** Prove that the centre of mass of a uniform right square-based pyramid of height h lies a distance $\dfrac{h}{4}$ from its base. **(6 marks)**

A solid is composed of a cube of side 8 cm joined at its top plane face to the base of a uniform square-based pyramid of side 8 cm and perpendicular height 6 cm. The centres of their adjoining square faces coincide at the point O and their axes are collinear.

  b Given that the cube has twice the density of the pyramid, find the position of the centre of mass of the composite body. **(4 marks)**

(E/P) **4 a** Find the distance of the centre of mass of a solid tetrahedron from its base, giving your answer in terms of the height of the tetrahedron, h. **(6 marks)**

A solid is composed of two solid tetrahedrons of side 9 cm joined together. The centres of their adjoining faces coincide at the point O and their axes are collinear.

  b Given that the bottom tetrahedron has three times the density of the top tetrahedron, find the position of the centre of mass of the composite body. **(4 marks)**

(E/P) **5** A truncated square-based pyramid has a height of 5 cm and a base length 10 cm. The top face of the truncated square-based pyramid has an area of 25 cm².

  a Find the height of the centre of mass of the truncated square-based pyramid above its base. **(6 marks)**

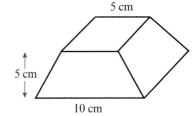

The truncated square-based pyramid and a cube of side length 5 cm are joined at the smaller square face of the truncated pyramid. The centres of their adjoining square faces coincide at the point O and their axes are collinear.

  b Given that the cube has twice the density of the truncated square-based pyramid, find the distance of the centre of mass of the composite body from O. **(4 marks)**

(E/P) **6** A solid wooden bowl is modelled as a uniform solid right circular cylinder with height $2r$ and radius $4r$. The centre of one face is at O. A hemisphere of diameter $2r$ is removed from the bowl. The centre of the hemisphere lies on a diameter of the circular face of the cylinder, and the point O lies on the circumference of the circular face of the hemisphere.

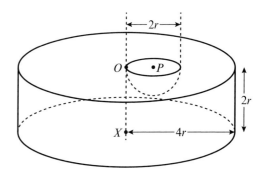

  a Show that the centre of mass of the whole bowl is at a vertical distance $\dfrac{381r}{376}$ from the plane face containing O. **(6 marks)**

The hemisphere is filled with water. The density of the water is twice the density of the wood from which the bowl is made.

b Find

  i the vertical distance of the centre of mass of the filled bowl from the plane face containing O

  ii the horizontal distance from the axis of the cylinder, OX, to the centre of mass of the filled bowl.

In each case, give your answer in the form kr where k is a rational number to be found. **(6 marks)**

Chapter review (5)   **SKILLS**   PROBLEM-SOLVING

  1 The curve shows a sketch of the region R bounded by the curve with equation $y^2 = 4x$ and the line with equation $x = 4$. The unit of length on both the axes is the centimetre. The region R is rotated through π radians about the x-axis to form a solid S.

  a Show that the volume of the solid S is $32\pi\,\text{cm}^3$.

                                        (4 marks)

  b Given that the solid is uniform, find the distance of the centre of mass of S from O.

                                        (6 marks)

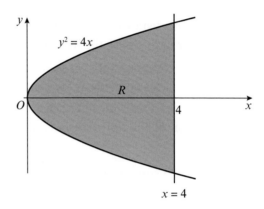

E/P **2** The region R is bounded by the curve with equation $y = \frac{1}{x}$, the lines $x = 1$, $x = 2$ and the x-axis, as shown in the diagram. The unit of length on both the axes is $1\,\text{m}$. A solid plinth is made by rotating R through 2π radians about the x-axis.

  a Show that the volume of the plinth is $\frac{\pi}{2}\,\text{m}^3$.

                                        (4 marks)

  b Find the distance of the centre of mass of the plinth from its larger plane face, giving your answer in cm to the nearest cm.

                                        (6 marks)

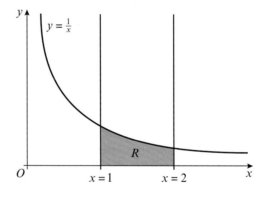

3 The diagram shows a cross-section of a uniform solid standing on horizontal ground. The solid consists of a uniform solid right circular cylinder, of diameter $80\,\text{cm}$ and height $40\,\text{cm}$, joined to a uniform solid hemisphere of the same density. The circular base of the hemisphere coincides with the upper circular end of the cylinder and has the same diameter as that of the cylinder. Find the distance of the centre of mass of the solid from the ground.

E/P

                                        (6 marks)

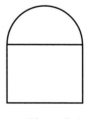

4 A simple wooden model of a rocket is made by taking a uniform cylinder, of radius r and height $3r$, and carving away part of the top two-thirds to form a uniform cone of height $2r$ as shown in cross-section in the diagram. Find the distance of the centre of mass of the model from its plane face.

(6 marks)

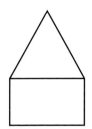

5 The finite region bounded by the curve $y = \sin x$, $0 \leqslant x \leqslant \dfrac{\pi}{6}$ and the x-axis is rotated through 2π radians about the x-axis to form a solid of revolution. Find the distance of its centre of mass from the y-axis, giving your answer to 3 significant figures.

6 Find, by integration, the coordinates of the centre of mass of the solid formed when the finite region bounded by $y^2 = \ln x$, the lines $x = 1$, $x = 2$ and the x-axis is rotated through 2π radians about the x-axis.

Challenge

A finite region is bounded by the curve with equation $y = \cos x$, the positive x-axis and the positive y-axis in the region $0 \leq x \leq \frac{\pi}{2}$

This region is rotated by 2π radians about the x-axis to form a uniform solid.

Work out the x-coordinate of the centre of mass.

Summary of key points

1 **The centre of mass of a uniform lamina** may be found using the formulae

$$\bullet \quad \overline{x} = \frac{\displaystyle\int_a^b xy\,\mathrm{d}x}{\displaystyle\int_a^b y\,\mathrm{d}x} \text{ and } \overline{y} = \frac{\displaystyle\int_a^b \tfrac{1}{2}y^2\,\mathrm{d}x}{\displaystyle\int_a^b y\,\mathrm{d}x}$$

$$\bullet \quad M\overline{x} = \int_a^b \rho xy\,\mathrm{d}x \text{ and } M\overline{y} = \int_a^b \frac{1}{2}\rho y^2\,\mathrm{d}x$$

where $M = \displaystyle\int_a^b \rho y\,\mathrm{d}x$, and is the total mass of the lamina, and ρ is the mass per unit area of the lamina.

2 **Standard results for uniform laminas and for arcs**

Lamina	Centre of mass along axis of symmetry
Semicircle, radius r	$\dfrac{4r}{3\pi}$ from the centre
Sector of circle, radius r, angle at centre 2α	$\dfrac{2r\sin\alpha}{3\alpha}$ from the centre
Circular arc, radius r, angle at centre 2α	$\dfrac{r\sin\alpha}{\alpha}$ from the centre

3 For a uniform solid body
- the weight is evenly distributed through the body
- the centre of mass will lie on any axis of symmetry
- the centre of mass will lie on any plane of symmetry

4 From symmetry, the centre of mass of some uniform bodies is at their geometric centre. These include the cube, the cuboid, the sphere, the right circular cylinder and the right circular prism.

5 For a **uniform solid of revolution**, where the revolution is about the x-axis, the centre of mass lies on the x-axis, by symmetry, and its position on the axis is given by the formulae

$$\bar{x} = \frac{\int \pi y^2 x \, dx}{\int \pi y^2 \, dx} \text{ or } M\bar{x} = \int \rho \pi y^2 x \, dx$$

where M is the known mass of the solid and ρ is its density.

6 For a **uniform solid of revolution**, where the revolution is about the y-axis, the centre of mass lies on the y-axis, by symmetry, and its position on the axis is given by the formulae

$$\bar{y} = \frac{\int \pi x^2 y \, dy}{\int \pi x^2 \, dy} \text{ or } M\bar{y} = \int \rho \pi x^2 y \, dy$$

where M is the known mass of the solid and ρ is its density.

7 Standard results for uniform bodies

Body	Centre of mass along axis of symmetry
Solid hemisphere, radius R	$\frac{3}{8}R$ from the centre
Hemispherical shell, radius R	$\frac{1}{2}R$ from the centre
Solid right circular cone, height h	$\frac{3}{4}h$ from the vertex, or $\frac{1}{4}h$ from the circular base
Conical shell, height h	$\frac{2}{3}h$ from the vertex, or $\frac{1}{3}h$ from the circular base

6 STATICS OF RIGID BODIES

6.1

Learning objectives

After completing this chapter you should be able to:

● Solve rigid body in equilibrium problems → **pages 148–153**

● Determine whether a rigid body on an inclined plane will slide
  or topple → **pages 153–163**

Prior knowledge check

1 A particle of mass m is suspended by two inelastic
light strings as shown.

Find T_1 and T_2 in terms of m and g.

← **Mechanics 1 Section 7.2**

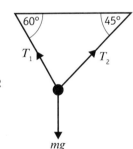

2 A particle is acted upon by 3 coplanar
horizontal forces as shown in the diagram.

a Resolve these forces in the directions
North and East.

b Hence determine the magnitude of
the resultant of these forces.

← **Mechanics 1 Section 7.2**

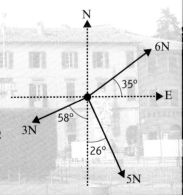

For a structure such as a
building to stay balanced,
the line of action of its own
weight, from its centre of
mass, must lie inside the
region where the shape
makes contact with the
ground.

6.1 Rigid bodies in equilibrium

You can solve problems about rigid bodies which are suspended in equilibrium.

Links You can combine resultant forces and moments to solve problems involving ladders leaning against walls. ← Mechanics 2 Section 6.1

If a body is resting in equilibrium then there is zero resultant force in any direction. This means that the sum of the components of all the forces in any direction is zero, and the sum of the moments of the forces about any point is zero.

- When a lamina is suspended freely from a fixed point or when it pivots freely about a horizontal axis, it will rest in equilibrium in a vertical plane with its centre of mass vertically below the point of suspension or pivot.

This result is also true for a rigid body.

Let the body be suspended from a point A. The body rests in equilibrium and the only forces acting on the body are its weight, W, and the force at point A, P. This implies that the forces must be equal and opposite and act in the same vertical line.

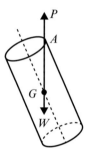

- When a rigid body is suspended freely from a fixed point and rests in equilibrium then its centre of mass is vertically below the point of suspension.

Example **1** **SKILLS** PROBLEM-SOLVING

A uniform solid hemisphere has radius r. It is suspended by a string attached to a point A on the rim of its base. Find the angle between the axis of the hemisphere and the downward vertical when the hemisphere is in equilibrium.

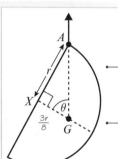

Draw a diagram showing the centre of mass, G, of the hemisphere below the point of suspension A.

Mark the angle between GA and the axis of the hemisphere and mark the radius and length XG.

The distance from the centre of mass to the base is $\dfrac{3r}{8}$ so $XG = \dfrac{3r}{8}$

Let $\angle XGA$ be θ

Use trigonometry to solve the problem.

Then $\tan \theta = \dfrac{r}{\dfrac{3}{8}r} = \dfrac{8}{3}$

So the required angle is 69.4° (3 s.f.)

Example **2** **SKILLS** CRITICAL THINKING

The diagram shows a uniform solid right circular cone, of mass M with radius r and height $3r$. P and Q are points at opposite ends of a diameter on the circular plane face of the cone. The cone is suspended from a horizontal beam by two vertical inelastic strings fastened at P and Q. Given that the cone is in equilibrium with its axis at an angle of 45° to the horizontal, find the values of the tensions in the strings, giving your answers in terms of M and g.

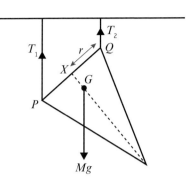

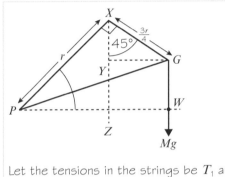

Let the tensions in the strings be T_1 and T_2.

Resolve vertically:

$$T_1 + T_2 = Mg \qquad (1)$$

Take moments about point P:

$$T_2 \times 2r \cos 45° = Mg\left(r \cos 45° + \frac{3r}{4} \sin 45°\right) \qquad (2)$$

Divide equation (2) through by $2r \cos 45°$

then $T_2 = \dfrac{7}{8} Mg$

Substitute into equation **(1)** to give $T_1 = \dfrac{1}{8} Mg$

So the tensions are $\dfrac{1}{8} Mg$ and $\dfrac{7}{8} Mg$

Moments could be taken about a number of points, but choosing point P eliminates T_1 and simplifies the algebra.

The distance PW is $PZ + ZW$.
Also $ZW = YG$.
So from the diagram
$$PW = r \cos 45° + \frac{3r}{4} \sin 45°$$

Example **3** **SKILLS** INTERPRETATION

Two smooth uniform spheres of radius 4 cm and mass 5 kg are suspended from the same point A by identical light inextensible strings of length 8 cm attached to their surfaces. The spheres hang in equilibrium, touching each other. What is the reaction between them?

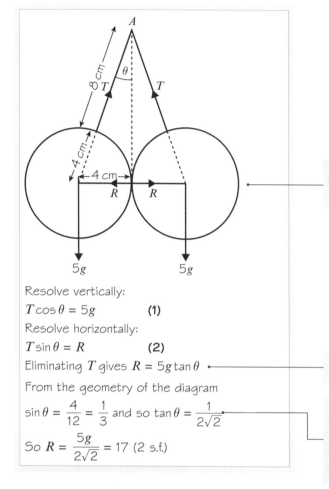

From symmetry, the tensions in the two strings are equal.

Resolve vertically:

$T \cos \theta = 5g$ **(1)**

Resolve horizontally:

$T \sin \theta = R$ **(2)**

Eliminating T gives $R = 5g \tan \theta$ Divide equation **(2)** by equation **(1)** to eliminate T.

From the geometry of the diagram

$\sin \theta = \dfrac{4}{12} = \dfrac{1}{3}$ and so $\tan \theta = \dfrac{1}{2\sqrt{2}}$

So $R = \dfrac{5g}{2\sqrt{2}} = 17$ (2 s.f.)

As $\sin \theta = \dfrac{1}{3}$ the triangle XYZ can be used to find XY using Pythagoras' theorem, and then to find $\tan \theta$

$XY^2 = 3^2 - 1^2$ $\therefore XY = \sqrt{8} = 2\sqrt{2}$

Exercise 6A   **SKILLS**   PROBLEM-SOLVING

1 A uniform solid right circular cone is suspended by a string attached to a point on the rim of its base. Given that the radius of the base is 5 cm and the height of the cone is 8 cm, find the angle between the vertical and the axis of the cone when it is in equilibrium.

2 A uniform solid cylinder is suspended by a string attached to a point on the rim of its base. Given that the radius of the base is 6 cm and the height of the cylinder is 10 cm, find the angle between the vertical and the circular base of the cylinder when it is in equilibrium.

3 A uniform hemispherical shell is suspended by a string attached to a point on the rim of its base. Find the angle between the vertical and the axis of the hemisphere when it is in equilibrium.

                                                                     (8 marks)

4 a Find the position of the centre of mass of the frustum of a uniform solid right circular cone, of end radii 4 cm and 5 cm and of height 6 cm. (Give your answer to 3 s.f.) **(8 marks)**

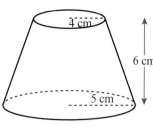

This frustum is now suspended by a string attached to a point on the rim of its smaller circular face, and hangs in equilibrium.

b Find the angle between the vertical and the axis of the frustum. **(3 marks)**

5 The diagram shows a uniform solid cylinder of mass 2 kg, height 2 m and radius 0.5 m. The cylinder is suspended from two light, inelastic, vertical strings attached to the upper rim of the cylinder at points A and B. The line AB forms a diameter of the top face of the cylinder and is inclined at an angle of 10° to the horizontal.

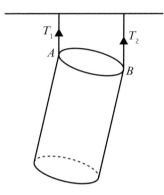

Given that the cylinder is in equilibrium, work out the tension in each string. **(5 marks)**

 6 A solid is formed by removing a solid hemisphere of diameter 6 cm and centre O from a solid, uniform right circular cylinder of height 10 cm and diameter 10 cm. The point O lies at the centre of one circular face of the cylinder.

The cylinder is suspended by two light, inextensible strings, which hang vertically. The strings are attached to points A and B on the rims of the solid, such that AB is parallel to the axis of the cylinder.

The solid has a density of $\frac{1}{\pi}$ kg m³, and hangs in equilibrium with AB horizontal.

a Find the tensions in the strings. **(10 marks)**

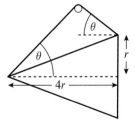

The string at B breaks, and the solid hangs in equilibrium from point A.

b Find the angle between the circular plane face of the solid and the vertical. **(3 marks)**

E/P **7** A uniform solid right circular cone has base radius r, height $4r$ and mass m kg. One end of a light inextensible string is attached to the vertex of the cone and the other end is attached to a point on the rim of the base. The string passes over a smooth peg and the cone rests in equilibrium with the axis horizontal, and with the strings equally inclined to the horizontal at an angle θ, as shown in the diagram.

a Show that angle θ satisfies the equation $\tan \theta = \frac{1}{2}$ **(4 marks)**

b Find the tension in the string, giving your answer as an exact multiple of mg. **(4 marks)**

(E/P) **8** A plastic mould is formed by removing a solid hemisphere of radius 40 cm and centre O from a solid hemisphere of metal of radius 60 cm and centre O, as shown in the diagram.

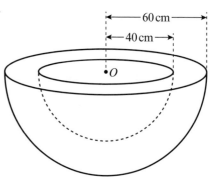

The mould is filled to the brim with molten plastic, which is allowed to solidify (i.e. to become solid). The metal used to form the mould has 10 times the density of the plastic.

After the mould is filled and set, it is suspended from a point on the outer rim of its plane face, and hangs in equilibrium.

Find the angle that the plane face makes with the vertical. **(10 marks)**

6.2 Toppling and sliding

You can determine whether a body will remain in equilibrium or if equilibrium will be broken by sliding or toppling.

■ **If a body rests in equilibrium on a rough inclined plane, then the line of action of the weight of the body must pass through the face of the body which is in contact with the plane.**

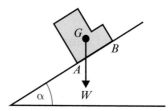

In the diagram, the weight of the body produces a clockwise moment about A which keeps it in contact with the plane.

If the angle of the plane is increased so that the line of action of the weight passes outside the face AB then the weight produces an anticlockwise moment about A. If the plane is sufficiently rough to preventy sliding, then the body will **topple** over.

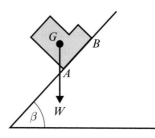

 Online Explore toppling and sliding using GeoGebra.

The only forces acting on the body are its weight and the total reaction between the plane and the body. The total reaction between the plane and the body consists of a normal reaction force and a friction force. As the body is in equilibrium, these forces must be equal and opposite and act in the same vertical line.

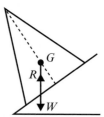

Example 4 SKILLS INTERPRETATION

A solid uniform cylinder of radius 3 cm and height 5 cm has a solid uniform hemisphere made from the same material, of radius 3 cm, joined to it so that the base of the hemisphere coincides with one circular end of the cylinder.

a Find the position of the centre of mass of the composite body.

The composite body is placed with the circular face of the cylinder on a rough inclined plane, which is inclined at an angle α to the horizontal. Given that the plane is sufficiently rough to prevent sliding,

b show that equilibrium is maintained provided that $\tan \alpha < \dfrac{28}{33}$

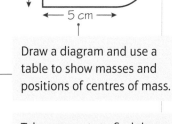

a	Shape	Mass	Ratio of masses	Height of centre of mass above base in cm
	Cylinder	$\rho\pi\, 3^2 \times 5 = 45\rho\pi$	5	2.5
	Hemisphere	$\frac{2}{3}\rho\pi\, 3^3 = 18\rho\pi$	2	$5 + \frac{3}{8} \times 3 = 6.125$
	Composite body	$63\,\rho\pi$	7	$\bar{x}$

Draw a diagram and use a table to show masses and positions of centres of mass.

So $5 \times 2.5 + 2 \times 6.125 = 7\bar{x}$

and $\bar{x} = \dfrac{24.75}{7} = \dfrac{99}{28}$

Take moments to find the position of the centre of mass of the composite body.

b From the diagram, $\tan \alpha = \dfrac{3}{\frac{99}{28}} = 3 \times \dfrac{28}{99} = \dfrac{28}{33}$

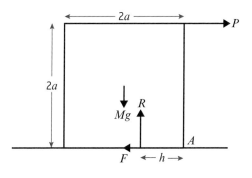

If α is smaller than this value, G is above a point of contact and equilibrium is maintained. Equilibrium is maintained provided that $\tan \alpha < \dfrac{28}{33}$

The limiting case is shown in the diagram, where the point vertically below the centre of mass, G, is on the edge of the area of contact.

If $\tan \alpha > \dfrac{28}{33}$ then the body will topple over because there will be a turning effect about the point A.

Example 5   **SKILLS**   CRITICAL THINKING

A uniform solid cube of mass M and side $2a$ rests on a rough horizontal plane. The coefficient of friction is μ. A horizontal force of magnitude P is applied at the midpoint of an upper edge, perpendicular to that edge, as shown in the diagram. The reaction between the plane and the cube comprises a normal reaction force R and a friction force F and acts at a distance h from a lower edge of the cube as shown in the diagram.

Find whether the cube remains in equilibrium, or whether the equilibrium is broken by sliding or toppling, in each of the following cases. Also determine the value of F and the value of h in each case, giving your answers in terms of M, g and a as appropriate

a $P = 0$ **b** $P = \dfrac{1}{2}Mg$ and $\mu = \dfrac{3}{4}$ **c** $P = \dfrac{1}{4}Mg$ and $\mu = \dfrac{1}{5}$

For equilibrium:

Resolve horizontally →

$\qquad P - F = 0$, so $F = P$ $\qquad\qquad\qquad$ (1)

Resolve vertically ↑

$\qquad R - Mg = 0$ so $R = Mg$ $\qquad\qquad\quad$ (2)

Take moments about point A:

$\qquad P \times 2a + R \times h = Mg \times a$ $\qquad\quad$ (3)

a When $P = 0$

$\qquad$ Substituting result from equation **(2)** into equation **(3)**:

$\qquad\qquad 0 + Mgh = Mga$

$\qquad\qquad h = a$

$\qquad$ Substituting $P = 0$ into equation **(1)** gives $F = 0$.

$\qquad$ The cube remains in equilibrium. It does not slide and does

$\qquad$ not topple.

b When $P = \dfrac{1}{2}Mg$

$\qquad$ Substituting result from equation **(2)** into equation **(3)**:

$\qquad\qquad \dfrac{1}{2}Mg \times 2a + Mgh = Mga$

$\qquad h = 0$, and the cube is about to topple.

$\qquad$ Substituting $P = \dfrac{1}{2}Mg$ into equation **(1)** gives $F = \dfrac{1}{2}Mg$

$\qquad$ But $\mu R = \dfrac{3}{4}Mg$ so $F < \mu R$ and the body does not slide.

c When $P = \dfrac{1}{4}Mg$

$\qquad$ Substituting result from equation **(2)** into equation **(3)**:

$\qquad\qquad \dfrac{1}{4}Mg \times 2a + Mgh = Mga$

$\qquad h = 1/2\, a$, and the cube does not topple.

$\qquad$ Substituting $P = \dfrac{1}{4}Mg$ into equation **(1)** would give $F = \dfrac{1}{4}Mg$

$\qquad$ But this is impossible as the maximum value that F can

$\qquad$ take is μR and $\mu R = \dfrac{1}{5}Mg$ so $F = \dfrac{1}{5}Mg$

$\qquad$ The cube will slide if the force P is maintained.

First use the conditions for equilibrium in the general case, i.e. resolve horizontally, resolve vertically and take moments.

A is the point on the bottom edge of the cube shown in the diagram.

For part **a** substitute $P = 0$ in equations **(1)**, **(2)** and **(3)**.

In part **a** the normal reaction acts at the centre of the base.

For part **b** substitute $P = \dfrac{1}{2}mg$ in equations **(1)**, **(2)** and **(3)**.

In part **b** the normal reaction acts at A and so toppling is about to occur around the edge through A.

This implies that the body does not topple as the reaction is within the area of the base.

Assuming equilibrium leads to a contradiction. This cube will slide as the force P exceeds the maximum friction force.

- A body is on the point of sliding when $F = \mu R$

- A body is on the point of toppling when the reaction acts at the point about which turning can take place.

Example 6 **SKILLS** PROBLEM-SOLVING

A uniform solid cube of mass M and side $2a$, is placed on a rough inclined plane which is at an angle α to the horizontal, where $\tan \alpha = \dfrac{1}{2}$ The coefficient of friction is μ.

Show that if $\mu < \dfrac{1}{2}$ the cube will slide down the slope.

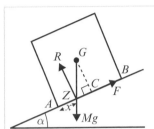

$R(\nwarrow)\ R - Mg\cos\alpha = 0$

$R = Mg\cos\alpha$

$R(\nearrow)\ F - Mg\sin\alpha = 0$

$F = Mg\sin\alpha$

Draw a diagram showing G the centre of mass of the cube, and points A and B on the edges of the cube in the same vertical plane as G.

First use the conditions for equilibrium, i.e. resolve along and perpendicular to the plane and use $F \leqslant \mu R$, where F is the force of friction and R is the normal reaction.

For equilibrium $F < \mu R$, i.e. $\mu > \dfrac{F}{R}$

So $\mu > \dfrac{Mg\sin\alpha}{Mg\cos\alpha}$

i.e. $\mu > \tan\alpha$ and so $\mu > \dfrac{1}{2}$ for equilibrium.

So if $\mu < 1/2$ the cube will slide.

Let the point Z, vertically below the centre of mass G, be at a distance x from A up the plane.

Let C be the midpoint of AB.

From triangle GCZ, $\dfrac{a - x}{a} = \tan\alpha = \dfrac{1}{2}$

So $2a - 2x = a$

$x = \dfrac{1}{2}a$

So Z is between A and B and the cube does not topple.

As $\tan\alpha = \dfrac{1}{2}$ the limiting case is when $\mu = \dfrac{1}{2}$

Draw an enlarged triangle if it helps.

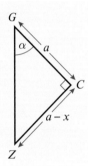

The normal reaction would act at point Z, which could also be established by taking moments and using the equilibrium conditions.

So the weight of the cube acts through a point within the area of contact.

Example 7   **SKILLS**   PROBLEM-SOLVING

The cube in the previous example is now replaced by a cylinder of mass M with base radius a and height $6a$. The cylinder is placed on the rough inclined plane, which is inclined at an angle α to the horizontal, where $\tan \alpha = \frac{1}{2}$ The coefficient of friction is μ. Show that if $\mu > \frac{1}{2}$ the cylinder will topple about the lower edge.

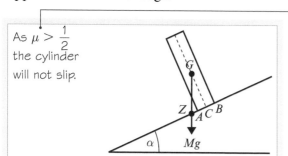

As $\mu > \frac{1}{2}$
the cylinder
will not slip.

Let the point Z, vertically below the centre of mass G, be at a distance x from A down the plane.

Let C be the midpoint of AB.

From triangle GCZ, $\dfrac{a + x}{3a} = \tan \alpha = \dfrac{1}{2}$

So $2a + 2x = 3a$

$x = \dfrac{1}{2}a$

So Z is not between A and B and the cylinder will topple.

Resolve along and perpendicular to the plane and use $F \leqslant \mu R$

Draw an enlarged triangle GZC.

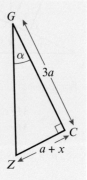

Another method you could use is to let the reaction act at point A and show that even in this position there will be a turning moment about A and therefore no equilibrium.

Exercise 6B   **SKILLS**   PROBLEM-SOLVING

1   A uniform solid cylinder with base diameter 4 cm stands on a rough plane inclined at 40° to the horizontal. Given that the cylinder does not topple, find the maximum possible height of the cylinder.

2   A uniform solid right circular cylinder with base radius 3 cm and height 10 cm is placed with its circular plane base on a rough plane. The plane is gradually tilted until the cylinder topples over. Given that the cylinder does not slide down the plane before it topples,

   a find the angle that the plane makes with the horizontal at the point where the cylinder topples

   b find the minimum possible value of the coefficient of friction between the cylinder and the plane.

P   **3**   A uniform solid right circular cone with base radius 5 cm and height h cm is placed with its circular plane base on a rough plane. The coefficient of friction is $\frac{\sqrt{3}}{3}$

The plane is gradually tilted.

When the plane makes an angle of θ with the horizontal, the cone is about to slide and topple at the same time. Find

   a θ

   b h

P   **4**   A uniform solid right circular cone of mass M with base radius r and height $2r$ is placed with its circular plane base on a rough horizontal plane. A force P is applied to the vertex V of the cone at an angle of 60° above the horizontal as shown in the diagram.

The cone begins to topple and to slide at the same time.

   a   Find the magnitude of the force P in terms of M.

   b   Calculate the value of the coefficient of friction.

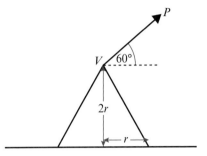

P   **5**   A frustum of a right circular solid cone has two plane circular end faces with radii r and $2r$ respectively. The distance between the end faces is $2r$.

   a   Show that the centre of mass of the frustum is at a distance $\frac{11r}{14}$ from the larger circular face.

   b   Find whether this solid can rest without toppling on a rough plane, inclined to the horizontal at an angle of 40°, if the face in contact with the inclined plane is

     i   the large circular end   **ii**   the small circular end.

   c   In order to answer part **b** you assumed that slipping did not occur.

     What does this imply about the coefficient of friction μ?

P   **6**   A uniform cube with edges of length $6a$ and weight W stands on a rough horizontal plane. The coefficient of friction is μ. A gradually increasing force P is applied at right angles to a vertical face of the cube at a point which is a distance a above the centre of that face.

   a   Show that equilibrium will be broken by sliding or toppling depending on whether

     $\mu < \frac{3}{4}$ or $\mu > \frac{3}{4}$

   b   If $\mu = \frac{1}{4}$, and the cube is about to slip, find the distance from the point where the normal reaction acts, to the nearest vertical face of the cube.

E/P **7** Two uniform, rough, solid cuboids of dimensions 20 cm × 20 cm × 30 cm are stacked as shown in the diagram shown opposite. Cuboid A has density ρ and cuboid B has density $k\rho$, for some constant k.

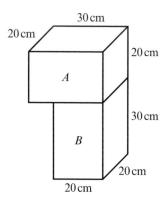

The cuboids are placed on a rough plane inclined at an angle of θ to the horizontal, as shown in the diagram opposite. The line of greatest slope of the plane is parallel to the plane faces shown in the diagram, and the angle of inclination of the plane is gradually increased.

a In the case when $k = 5$, find the angle of inclination at which the whole body begins to topple. **(8 marks)**

b Given that cuboid A topples off cuboid B before the whole body topples, find the range of possible values of k. **(4 marks)**

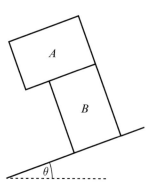

E/P **8** A uniform solid paperweight is in the shape of a frustum of a cone. It is formed by removing a right circular cone of height h from a right circular cone of height $2h$ and base radius $2r$.

a Show that the centre of mass of the paperweight lies at a height of $\dfrac{11}{28}h$ from its base. **(6 marks)**

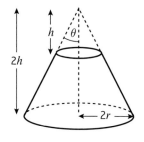

When placed with its curved surface on a horizontal plane, the paperweight is on the point of toppling.

b Find θ, the semi-vertical angle of the cone, to the nearest degree. **(4 marks)**

E/P **9** A uniform solid cone of mass M, height 8 cm and radius 3 cm, is placed with its circular base on a horizontal plane. The coefficient of friction is μ. A horizontal force of magnitude P is applied at the vertex of the cone.

a Find the value of P which will cause the cone to slide giving your answer in terms of μ, M and g. **(4 marks)**

b Find the value of P which will cause the cone to topple, giving your answer in terms of M and g. **(2 marks)**

c State whether the cone will slide or topple if

i $\mu = \dfrac{1}{4}$ **(1 mark)**

ii $\mu = \dfrac{1}{2}$ **(1 mark)**

iii $\mu = \dfrac{3}{8}$ **(1 mark)**

(E/P) **10** A uniform solid cylinder of mass 200 g, height 4 cm and radius 3 cm rests with its circular base on a plane. The plane makes an angle α with the horizontal where $\tan\alpha = \dfrac{3}{4}$

The coefficient of friction is $\dfrac{6}{17}$

A horizontal force P is applied to the highest point of the cylinder as shown in the diagram.

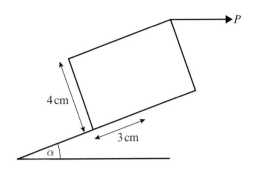

a Find the value of P which will just cause the cylinder to topple about the highest point of the base. **(4 marks)**

b Find the value of P which would cause the cylinder to slide up the plane. **(6 marks)**

c Show that the cylinder topples before it slides. **(1 mark)**

Chapter review **6** **SKILLS** PROBLEM-SOLVING

(E/P) **1** The diagram shows a cross-section containing the axis of symmetry of a uniform body consisting of a solid right circular cylinder of base radius r and height kr surmounted by a solid hemisphere of radius r.

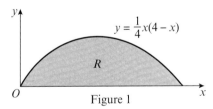

a Given that the centre of mass of the body is at the centre C of the common face of the cylinder and the hemisphere, find the value of k, giving your answer to 2 significant figures. **(5 marks)**

b Explain briefly why the body remains at rest when it is placed with any part of its hemispherical surface in contact with a horizontal plane. **(1 mark)**

(E/P) **2** A uniform lamina occupies the region R bounded by the x-axis and the curve with equation

$y = \dfrac{1}{4}x(4-x)$, $0 \leqslant x \leqslant 4$, as shown in Figure 1.

a Show by integration that the y-coordinate of the centre of mass of the lamina is $\dfrac{2}{5}$ **(8 marks)**

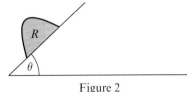

Figure 1

A uniform prism P has cross-section R. The prism is placed with its rectangular face on a slope inclined at an angle θ to the horizontal. The cross-section R lies in a vertical plane as shown in Figure 2. The surfaces are sufficiently rough to prevent P from sliding.

Figure 2

b Find the angle θ, for which P is about to topple. **(2 marks)**

3 A uniform semicircular lamina has radius $2a$ and the midpoint of the bounding diameter AB is O.

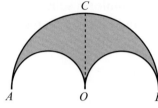

a Using integration, show that the centre of mass of the lamina is at a distance $\frac{8a}{3\pi}$ from O. **(8 marks)**

The two semicircular laminas, each of radius a and with AO and OB as diameters, are cut away from the original lamina to leave the lamina $AOBC$ shown in the diagram, where OC is perpendicular to AB.

b Show that the centre of mass of the lamina $AOBC$ is at a distance $\frac{4a}{\pi}$ from O. **(6 marks)**

The lamina $AOBC$ is of mass M and a particle of mass M is attached to the lamina at B to form a composite body.

c State the distance of the centre of mass of the body from OC and from OB. **(1 mark)**

The body is smoothly hinged at A to a fixed point and rests in equilibrium in a vertical plane.

d Calculate, to the nearest degree, the acute angle between AB and the horizontal. **(3 marks)**

4 A uniform wooden 'mushroom', used in a game, is made by joining a solid cylinder to a solid hemisphere. They are joined symmetrically, such that the centre O of the plane face of the hemisphere coincides with the centre of one of the ends of the cylinder. The diagram shows the cross-section through a plane of symmetry of the mushroom, as it stands on a horizontal table.

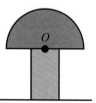

The radius of the cylinder is r, the radius of the hemisphere is $3r$, and the centre of mass of the mushroom is at the point O.

a Show that the height of the cylinder is $r\sqrt{\frac{81}{2}}$ **(6 marks)**

The table top, which is rough enough to prevent the mushroom from sliding, is slowly tilted until the mushroom is about to topple.

b Find, to the nearest degree, the angle with the horizontal through which the table top has been tilted. **(3 marks)**

5 Figure 1 shows a finite region A which is bounded by the curve with equation $y^2 = 4ax$, the line $x = a$ and the x-axis. A uniform solid S_1 is formed by rotating A through 2π radians about the x-axis.

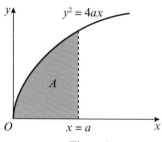

a Show that the volume of S_1 is $2\pi a^3$. **(3 marks)**

b Show that the centre of mass of S_1 is a distance $\frac{2a}{3}$ from the origin O. **(4 marks)**

Figure 1

Figure 2 shows a cross-section of a uniform solid S which has been obtained by attaching the plane base of solid S_1 to the plane base of a uniform hemisphere S_2 of base radius $2a$.

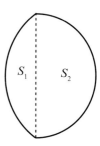

Figure 2

c Given that the densities of solids S_1 and S_2 are ρ_1 and ρ_2 respectively, find the ratio $\rho_1 : \rho_2$ which ensures that the centre of mass of S lies in the common plane face of S_1 and S_2. **(6 marks)**

d Given that $\rho_1 : \rho_2 = 6$, explain why the solid S may rest in equilibrium with any point of the curved surface of the hemisphere in contact with a horizontal plane. **(2 marks)**

 6 A mould for a right circular cone, base radius r and height h, is produced by making a conical hole in a uniform cylindrical block, base radius $2r$ and height $3r$.

The axis of symmetry of the conical hole coincides with that of the cylinder, and AB is a diameter of the top of the cylinder, as shown in the diagram.

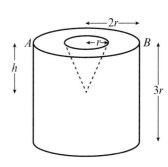

a Show that the distance from AB of the centre of mass of the mould is $\dfrac{216r^2 - h^2}{4(36r - h)}$ **(8 marks)**

The mould is suspended from the point A, and hangs freely in equilibrium.

b In the case $h = 2r$, calculate, to the nearest degree, the angle between AB and the downward vertical. **(4 marks)**

 7 A compound solid is made from an inverted frustum and a hemisphere as shown opposite.

The base of the frustum has a radius of $5\,\text{cm}$ and the hemisphere has a radius of $10\,\text{cm}$.

a Given that the density, ρ, of the hemisphere is three times the density of the frustum, find the height of the centre of mass of the compound solid from its base. **(8 marks)**

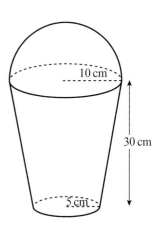

The compound solid is placed on a rough plane inclined at $45°$ to the horizontal.

b Given that the total mass of the compound solid is $m\,\text{kg}$, work out the minimum horizontal force, in terms of m, that must be applied at A to stop the solid from toppling over. The plane is sufficiently rough to prevent slipping. **(4 marks)**

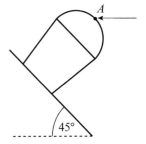

8 A solid is formed by rotating the region enclosed by the curve $y = \dfrac{2}{x+3}$, the lines $x = 2$ and $x = 4$ and the x-axis, though 360° around the x-axis.

The solid is then placed on its smaller circular end on rough horizontal ground.

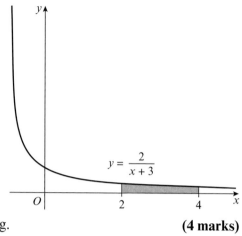

a Find the height of the centre of mass of the solid above the ground. **(10 marks)**

The solid is then placed on a rough inclined ramp angled at $\theta°$ to the horizontal.

b Assuming that the solid does not slip, work out the value of θ when the solid is on the point of tipping. **(4 marks)**

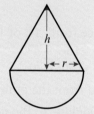
Summary of key points

1 A rigid body is in equilibrium if
  - there is zero resultant force in any direction, i.e. the sum of the components of all the forces in any direction is zero, and
  - the sum of the moments of the forces about any point is zero.

2 • **When a lamina is suspended freely** from a fixed point or pivots freely about a horizontal axis it will rest in equilibrium in a vertical plane with its centre of mass vertically below the point of suspension or pivot.
  - **When a rigid body is suspended freely** from a fixed point and rests in equilibrium then its centre of mass is vertically below the point of suspension.

3   If a body rests in equilibrium on a rough inclined plane, then the line of action of the weight of the body must pass through the face of the body which is in contact with the plane.

4   You can establish whether equilibrium will be broken by sliding or by toppling by considering
    • a body is on the point of sliding when $F = \mu R$
    • a body is on the point of toppling when the reaction acts at the point about which turning can take place.

Review exercise

2

Whenever a numerical value of g is required, take $g = 9.8 \text{ m s}^{-2}$

(E) **1** A circular flywheel of diameter 7 cm is rotating about the axis through its centre and perpendicular to its plane with constant angular speed 1000 revolutions per minute.

Find, in m s^{-1} to 3 significant figures, the speed of a point on the rim of the flywheel. **(2)**

← Mechanics 3 Section 4.1

(E) **2** A particle P of mass 0.5 kg is attached to one end of a light inextensible string of length 1.5 m. The other end of the string is attached to a fixed point A. The particle is moving, with the string taut, in a horizontal circle with centre O vertically below A. The particle is moving with constant angular speed 2.7 rad s^{-1}. Find

a the tension in the string **(4)**

b the angle, to the nearest degree, that AP makes with the downward vertical. **(4)**

← Mechanics 3 Section 4.2

(E/P) **3** A car moves round a bend which is banked at a constant angle of 10° to the horizontal. When the car is travelling at a constant speed of 18 m s^{-1}, there is no sideways frictional force on the car. The car is modelled as a particle moving in a horizontal circle of radius r metres. Calculate the value of r. **(6)**

← Mechanics 3 Section 4.3

(E) **4** A cyclist is travelling around a circular track which is banked at 25° to the horizontal. The coefficient of friction between the cycle's tyres and the track is 0.6. The cyclist moves with constant speed in a horizontal circle of radius 40 m, without the tyres slipping.

Find the maximum speed of the cyclist. **(9)**

← Mechanics 3 Section 4.3

(E) **5**

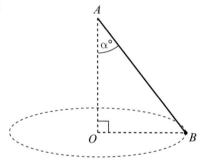

A metal ball B of mass m is attached to one end of a light inextensible string. The other end of the string is attached to a fixed point A. The ball B moves in a horizontal circle with centre O vertically below A, as shown in the diagram. The string makes a constant angle $\alpha°$ with the downward vertical and B moves with constant angular speed $\sqrt{2gk}$, where k is a constant. The tension in the string is $3mg$. By modelling B as a particle, find

a the value of α **(4)**

b the length of the string. **(5)**

← Mechanics 3 Section 4.3

6 A particle P of mass m moves on the smooth inner surface of a spherical bowl of internal radius r. The particle moves with constant angular speed in a horizontal circle, which is at a depth $\frac{1}{2}r$ below the centre of the bowl. Find

   a the normal reaction of the bowl on P **(4)**

   b the time for P to complete one revolution of its circular path. **(6)**

               ← **Mechanics 3 Section 4.3**

(E/P) 7

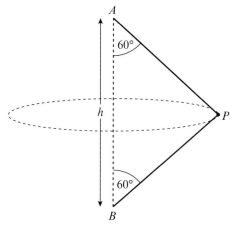

A particle P of mass m is attached to two light inextensible strings. The other ends of the string are attached to fixed points A and B. The point A is a distance h vertically above B. The system rotates about the line AB with constant angular speed ω. Both strings are taut and inclined at 60° to AB, as shown in the diagram. The particle moves in a circle of radius r.

   a Show that $r = \dfrac{\sqrt{3}}{2}h$ **(3)**

   b Find, in terms of m, g, h and ω, the tension in AP and the tension in BP. **(3)**

The time taken for P to complete one circle is T.

   c Show that $T < \pi\sqrt{\dfrac{2h}{g}}$ **(2)**

               ← **Mechanics 3 Section 4.3**

(E/P) 8

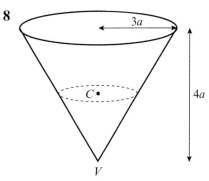

A hollow cone, of base radius $3a$ and height $4a$, is fixed with its axis vertical and vertex V downwards, as shown in the diagram. A particle moves in a horizontal circle with centre C, on the smooth inner surface of the cone with constant angular speed $\sqrt{\dfrac{8g}{9a}}$. Find the height of C above V. **(8)**

               ← **Mechanics 3 Section 4.3**

(E/P) 9 A rough disc rotates in a horizontal plane with constant angular velocity ω about a fixed vertical axis. A particle P of mass m lies on the disc at a distance $\frac{4}{3}a$ from the axis. The coefficient of friction between P and the disc is $\frac{3}{5}$. Given that P remains at rest relative to the disc,

   a prove that $\omega^2 \leqslant \dfrac{9g}{20a}$

The particle is now connected to the axis by a horizontal light elastic string of natural length a and modulus of elasticity $2mg$. The disc again rotates with constant angular velocity ω about the axis and P remains at rest relative to the disc at a distance $\frac{4}{3}a$ from the axis.

   b Find the greatest and least possible values of ω^2.

               ← **Mechanics 3 Section 4.3**

(E/P) 10 One end of a light inextensible string of length l is attached to a particle P of mass m. The other end is attached to a fixed point A. The particle is hanging freely at rest with the string vertical when it is projected horizontally with speed $\sqrt{\dfrac{5gl}{2}}$

    a Find the speed of P when the string is horizontal. **(4)**

When the string is horizontal it comes into contact with a small smooth fixed peg which is at the point B, where AB is horizontal, and $AB < l$

Given that the particle then describes a complete semi-circle with centre B,

    b find the least possible value of the length AB. **(9)**

← Mechanics 3 Section 1.4

(E/P) 11 One end of a light inextensible string of length l is attached to a fixed point A. The other end is attached to a particle P of mass m, which is hanging freely at rest at point B. The particle P is projected horizontally from B with speed $\sqrt{3gl}$

When AP makes an angle θ with the downward vertical and the string remains taut, the tension in the string is T.

    a Show that $T = mg(1 + 3\cos\theta)$ **(6)**

    b Find the speed of P at the instant when the string becomes slack. **(3)**

    c Find the maximum height above the level of B reached by P. **(5)**

← Mechanics 3 Section 4.4

(E/P) 12

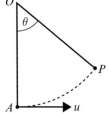

A particle of mass m is attached to one end of a light inextensible string of length l.

The other end of the string is attached to a fixed point O. The particle is hanging at the point A, which is vertically below O. It is projected horizontally with speed u. When the particle is at the point P, $\angle AOP = \theta$, as shown in the diagram. The string oscillates through an angle α on either side of OA where $\cos\alpha = \dfrac{2}{3}$

    a Find u in terms of g and l. **(3)**

When $\angle AOP = \theta$, the tension in the string is T.

    b Show that $T = \dfrac{mg}{3}(9\cos\theta - 4)$ **(4)**

    c Find the range of values of T during the oscillations. **(2)**

← Mechanics 3 Section 4.4

(E) 13 A smooth solid sphere, with centre O and radius a, is fixed to the upper surface of a horizontal table. A particle P is placed on the surface of the sphere at a point A, where OA makes an angle α with the upward vertical, and $0 < \alpha < \dfrac{\pi}{2}$

The particle is released from rest. When OP makes an angle θ with the upward vertical, and P is still on the surface of the sphere, the speed of P is v.

    a Show that $v^2 = 2ga(\cos\alpha - \cos\theta)$ **(4)**

Given that $\cos\alpha = \dfrac{3}{4}$, find

    b the value of θ when P loses contact with the sphere **(5)**

    c the speed of P as it hits the table. **(4)**

← Mechanics 3 Section 4.5

(E/P) 14

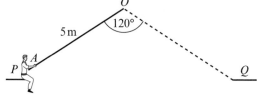

A trapeze artist of mass 60 kg is attached to the end A of a light inextensible rope OA of length 5 m. The artist must swing in an arc of a vertical circle, centre O, from a platform P to another platform Q, where PQ is horizontal. The other end of the rope is attached to the fixed point O which lies in the vertical plane containing PQ, with $\angle POQ = 120°$ and $OP = OQ = 5$ m, as shown in the diagram.

As part of her act, the artist projects herself from P with speed $\sqrt{15}$ m s^{-1} in a direction perpendicular to the rope OA and in the plane POQ. She moves in a circular arc towards Q. At the lowest point of her path she catches a ball of mass m kg which is travelling towards her with speed 3 m s^{-1} and parallel to QP. After catching the ball, she comes to rest at the point Q.

By modelling the artist and the ball as particles and ignoring her air resistance, find

a the speed of the artist immediately before she catches the ball   **(4)**

b the value of m **(6)**

c the tension in the rope immediately after she catches the ball.   **(2)**

← Mechanics 3 Section 4.4

(E) **15**

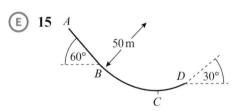

The diagram represents the path of a skier of mass 70 kg moving on a ski-slope $ABCD$. The path lies in a vertical plane. From A to B, the path is modelled as a straight line inclined at 60° to the horizontal. From B to D, the path is modelled as an arc of a vertical circle of radius 50 m. The lowest point of the arc BD is C.

At B, the skier is moving downwards with speed 20 m s^{-1}. At D, the path is inclined at 30° to the horizontal and the skier is

moving upwards. By modelling the slope as smooth and the skier as a particle, find

a the speed of the skier at C **(2)**

b the normal reaction of the slope on the skier at C **(2)**

c the speed of the skier at D **(2)**

d the change in the normal reaction of the slope on the skier as she passes B.   **(4)**

The model is refined to allow for the influence of friction on the motion of the skier.

e State briefly, with a reason, how the answer to part **b** would be affected by using such a model. (No further calculations are expected.)   **(1)**

← Mechanics 3 Section 4.4

(E) **16**

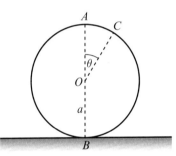

A particle is at the highest point A on the outer surface of a fixed smooth sphere of radius a and centre O. The lowest point B of the sphere is fixed to a horizontal plane. The particle is projected horizontally from A with speed u, where $u < \sqrt{(ag)}$. The particle leaves the sphere at the point C, where OC makes an angle θ with the upward vertical, as shown in the diagram above.

a Find an expression for $\cos \theta$ in terms of u, g and a.   **(7)**

The particle strikes the plane with speed $\sqrt{\dfrac{9ag}{2}}$

b Find, to the nearest degree, the value of θ.   **(7)**

← Mechanics 3 Section 4.5

(E) **17** A uniform solid is formed by rotating the region enclosed between the curve with equation $y = \sqrt{x}$, the x-axis and the line $x = 4$, through one complete revolution about the x-axis. Find the distance of the centre of mass of the solid from the origin O. **(5)**

← Mechanics 3 Section 5.1

(E) **18 a** Use integration to show that the centre of mass of a uniform semicircular lamina, of radius a, is a distance $\dfrac{4a}{3\pi}$ from the midpoint of its straight edge, O. **(4)**

A semicircular lamina, of radius b with O as the midpoint of its straight edge, is removed from the first lamina.

b Show that the centre of mass of the resulting lamina is at a distance $\bar{x}$ from O, where

$$\bar{x} = \frac{4}{3\pi} \frac{(a^2 + ab + b^2)}{(a + b)} \qquad \textbf{(6)}$$

c Hence find the position of the centre of mass of a uniform semicircular arc of radius a. **(2)**

← Mechanics 3 Section 5.2

(E/P) **19**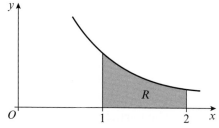

The shaded region R is bounded by the curve with equation $y = \dfrac{1}{2x^2}$, the x-axis and the lines $x = 1$ and $x = 2$, as shown above. The unit of length on each axis is 1 m. A uniform solid S has the shape made by rotating R through 360° about the x-axis.

a Show that the centre of mass of S is $\dfrac{2}{7}$ m from its larger plane face. **(8)**

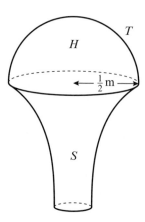

A sporting trophy T is a uniform solid hemisphere H joined to the solid S. The hemisphere has radius $\dfrac{1}{2}$ m and its plane face coincides with the larger plane face of S, as shown above. Both H and S are made of the same material.

b Find the distance of the centre of mass of T from its plane face. **(6)**

← Mechanics 3 Sections 5.1, 5.2

(E/P) **20 a** Show, by integration, that the centre of mass of a uniform solid hemisphere, of radius R, is at a distance $\dfrac{3}{8}R$ from its plane face.

The diagram shows a uniform solid top made from a right circular cone of base radius a and height ka and a hemisphere of radius a. The circular plane faces of the cone and hemisphere are coincident.

(6)

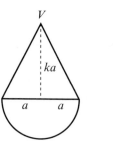

b Show that the distance of the centre of mass of the top from the vertex V of the cone is

$$\frac{(3k^2 + 8k + 3)a}{4(k + 2)} \qquad \textbf{(5)}$$

The manufacturer requires the top to have its centre of mass situated at the centre of the coincident plane faces.

c Find the value of k for this requirement. **(5)**

← Mechanics 3 Section 5.2

(E/P) **21** A bowl consists of a uniform solid metal hemisphere, of radius a and centre O, from which is removed the solid hemisphere of radius $\frac{1}{2}a$ with the same centre O.

a Show that the distance of the centre of mass of the bowl from O is $\frac{45}{112}a$ **(6)**

The bowl is fixed with its plane face uppermost and horizontal. It is now filled with liquid. The mass of the bowl is M and the mass of the liquid is kM, where k is a constant. Given that the distance of the centre of mass of the bowl and liquid together from O is $\frac{17}{48}a$

b find the value of k. **(6)**

← Mechanics 3 Section 5.3

(E/P) **22**

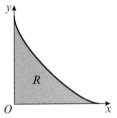

The shaded region R is bounded by part of the curve with equation $y = \frac{1}{2}(x-2)^2$, the x-axis and the y-axis, as shown above. The unit of length on both axes is 1 cm. A uniform solid S is made by rotating R through 360° about the x-axis. Using integration,

a calculate the volume of the solid S, leaving your answer in terms of π **(4)**

b show that the centre of mass of S is $\frac{1}{3}$ cm from its plane face. **(7)**

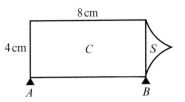

A tool is modelled as having two components, a solid uniform cylinder C and the solid S. The diameter of C is 4 cm and the length of C is 8 cm. One end of C coincides with the plane face of S. The components are made of different materials. The weight of C is $10W$ newtons and the weight of S is $2W$ newtons. The tool lies in equilibrium with its axis of symmetry horizontal on two smooth supports A and B, which are at the ends of the cylinder, as shown above.

c Find the magnitude of the force of the support A on the tool. **(5)**

← Mechanics 3 Sections 5.2, 6.1

(E) **23**

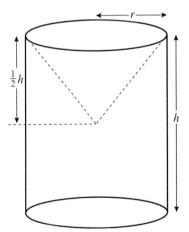

An ornament S is formed by removing a solid right circular cone, of radius r and height $\frac{1}{2}h$, from a solid uniform cylinder, of radius r and height h, as shown in the diagram.

a Show that the distance of the centre of mass of S from its plane face is $\frac{17}{40}h$ **(6)**

The ornament is suspended from a point on the circular rim of its open end. It hangs in equilibrium with its axis of symmetry inclined at an angle α to the horizontal. Given that $h = 4r$

b find, in degrees to one decimal place, the value of α. **(4)**

← Mechanics 3 Sections 5.2, 6.1

(E/P) **24** A closed container C consists of a thin uniform hollow hemispherical bowl of radius a, together with a lid. The lid is a thin uniform circular disc, also of radius a. The centre O of the disc coincides with the centre of the hemispherical bowl. The bowl and its lid are made of the same material.

a Show that the centre of mass of C is at a distance $\dfrac{1}{3}a$ from O. **(4)**

The container C has mass M. A particle of mass $\dfrac{1}{2}M$ is attached to the container at a point P on the circumference of the lid. The container is then placed with a point of its curved surface in contact with a horizontal plane. The container rests in equilibrium with P, O and the point of contact in the same vertical plane.

b Find, to the nearest degree, the angle made by the line PO with the horizontal. **(5)**

← Mechanics 3 Sections 5.2, 6.1

(E/P) **25** A uniform solid hemisphere H has base radius a and the centre of its plane circular face is C.

The plane face of a second hemisphere K, of radius $\dfrac{a}{2}$, and made of the same material as H, is stuck to the plane face of H, so that the centres of the two plane faces coincide at C, to form a uniform composite body S.

a Given that the mass of K is M, show that the mass of S is $9M$, and find, in terms of a, the distance of the centre of mass of the body S from C. **(5)**

A particle P, of mass M, is attached to a point on the edge of the circular face of H of the body S. The body S with P attached is placed with a point of the curved surface of the part H in contact with a horizontal plane and rests in equilibrium.

b Find the tangent of the acute angle made by the line PC with the horizontal. **(5)**

← Mechanics 3 Sections 5.2, 6.1

(E) **26**

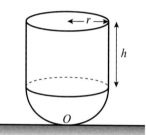

A child's toy consists of a uniform solid hemisphere attached to a uniform solid cylinder. The plane face of the hemisphere coincides with the plane face of the cylinder, as shown in the diagram above. The cylinder and the hemisphere each have radius r and the height of the cylinder is h. The material of the hemisphere is six times as dense as the material of the cylinder. The toy rests in equilibrium on a horizontal plane with the cylinder above the hemisphere and the axis of the cylinder vertical.

a Show that the distance d of the centre of mass of the toy from its lowest point O is given by

$$d = \frac{h^2 + 2hr + 5r^2}{2(h + 4r)} \qquad \textbf{(7)}$$

When the toy is placed with any point of the curved surface of the hemisphere resting on the plane it will remain in equilibrium.

b Find h in terms of r. **(3)**

← Mechanics 3 Sections 5.3, 6.1

(E) **27**

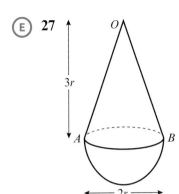

A child's toy consists of a uniform solid hemisphere, of mass M and base radius r, joined to a uniform solid right circular cone of mass m, where $2m < M$. The cone has vertex O, base radius r and height $3r$. Its plane face, with diameter AB, coincides with the plane face of the hemisphere, as shown in the diagram above.

a Show that the distance of the centre of mass of the toy from AB is

$$\frac{3(M - 2m)}{8(M + m)}r \qquad \textbf{(5)}$$

The toy is placed with OA on a horizontal surface. The toy is released from rest and does not remain in equilibrium.

b Show that $M > 26m$ **(4)**

← **Mechanics 3 Sections 5.3, 6.1**

(E) **28**

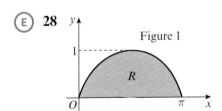

A uniform lamina occupies the region R bounded by the x-axis and the curve $y = \sin x$, $0 \leqslant x \leqslant \pi$, as shown in Figure 1.

a Show, by integration, that the y-coordinate of the centre of mass of the lamina is $\dfrac{\pi}{8}$ **(6)**

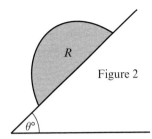

Figure 2

A uniform prism S has cross section R. The prism is placed with its rectangular face on a table which is inclined at an angle θ to the horizontal. The cross section R lies in a vertical plane as shown in Figure 2. The table is sufficiently rough to prevent S sliding. Given that S does not topple,

b find the largest possible value of θ. **(3)**

← **Mechanics 3 Sections 5.2, 6.2**

(E/P) **29**

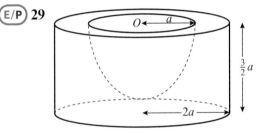

A uniform solid cylinder has radius $2a$ and height $\dfrac{3}{2}a$. A hemisphere of radius a is removed from the cylinder. The plane face of the hemisphere coincides with the upper plane face of the cylinder, and the centre O of the hemisphere is also the centre of this plane face, as shown in the diagram above. The remaining solid is S.

a Find the distance of the centre of mass of S from O. **(6)**

The lower plane face of S rests in equilibrium on a desk lid which is inclined at an angle θ to the horizontal. Assuming that the lid is sufficiently rough to prevent S from slipping, and that S is on the point of toppling when $\theta = \alpha$,

b find the value of α. **(3)**

Given instead that the coefficient of friction between S and the lid is 0.8, and that S is on the point of sliding down the lid when $\theta = \beta$,

c find the value of β. **(3)**

← **Mechanics 3 Sections 5.2, 6.2**

E/P **30**

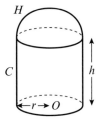

A body consists of a uniform solid circular cylinder C, together with a uniform solid hemisphere H which is attached to C. The plane face of H coincides with the upper plane face of C, as shown in the diagram. The cylinder C has base radius r, height h and mass $3M$. The mass of H is $2M$. The point O is the centre of the base of C.

a Show that the distance of the centre of mass of the body from O is

$$\frac{14h + 3r}{20}$$ **(5)**

The body is placed with its plane face on a rough plane which is inclined at an angle α to the horizontal, where $\tan \alpha = \dfrac{4}{3}$

The plane is sufficiently rough to prevent slipping. Given that the solid is on the point of toppling,

b find h in terms of r. **(4)**

← **Mechanics 3 Sections 5.3, 6.2**

E/P **31**

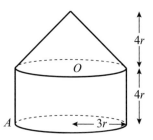

A toy is formed by joining a uniform solid right circular cone, of base radius $3r$ and height $4r$, to a uniform solid cylinder, also of radius $3r$ and height $4r$. The cone and cylinder are made from different materials, and the density of the cone is three times the density of the cylinder. The plane face of the cone coincides with a plane face of the cylinder, as shown in the diagram. The centre of this plane face is O.

a Find the distance of the centre of mass of the toy from O. **(6)**

The point A lies on the edge of the plane face of the cylinder which forms the base of the toy. The toy is suspended from A and hangs in equilibrium.

b Find, in degrees to one decimal place, the angle between the axis of symmetry of the toy and the vertical. **(4)**

The toy is placed with the curved surface of the cone on horizontal ground.

c Determine whether the toy will topple. **(4)**

← **Mechanics 3 Sections 5.3, 6.1, 6.2**

Challenge

1 The diagram shows two beads P_1 and P_2, of masses $5m$ and $7m$ respectively, threaded onto a smooth circular horizontal ring. The beads are projected in opposite directions at the same speed.

The beads collide with coefficient of restitution e at the point A, then collide again at the point B, where $\angle AOB = 90°$, as shown in the diagram.

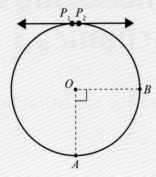

Find the value of e.

← Mechanics 2 Section 5
← Mechanics 3 Section 4.2

2 A roulette wheel is modelled as a circular ring of radius R resting on a smooth horizontal surface. A ball of mass m is held against the ring, and projected tangentially along the inside of the ring with initial speed u. The coefficient of friction between the ball and the ring is μ.

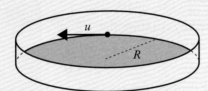

a Show that t seconds after the ball is projected its speed is $\dfrac{uR}{R + u\mu t}$

Given that $R = 0.5$ m, $u = 40$ m s^{-1} and $\mu = 0.25$

b find the time taken for the ball to complete its first complete revolution of the ring.

← Mechanics 3 Section 4.2

Exam practice

Mathematics
International Advanced Subsidiary/
Advanced Level Mechanics 3

Time: 1 hour 30 minutes
You must have: Mathematical Formulae and Statistical Tables, Calculator
Answer ALL questions

1   A particle P of mass $1.5\,\text{kg}$ is moving along the x-axis in the positive direction.
    At time $t = 0$, P passes through the origin with speed $8\,\text{ms}^{-1}$, and at time t seconds
    the resultant force acting on P is $4x\,\text{N}$, acting in the negative x direction.

    a Find v^2 in terms of x. **(6)**

    b How far does P travel before stopping? **(2)**

2   Figure 1 shows a sketch of the region R bounded by the
    curve with equation $y = \cos x$, the line $x = \dfrac{\pi}{4}$, the x-axis
    and the y-axis. A uniform solid S is formed by rotating the
    region R through 2π radians about the x-axis.

    a Show that the volume of S is $\dfrac{\pi}{8}(\pi + 2)$. **(4)**

    b Find, in terms of π, the distance of the centre of mass
    of S from O. **(7)**

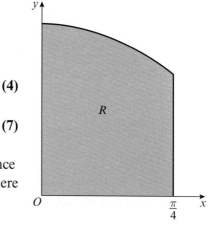

3   A particle P of mass m is above the Earth's surface at a distance
    x from the centre of the Earth, the Earth is modelled as a sphere
    of radius R and exerts a gravitational force on P which is
    inversely proportional to x^2.

    a Show that the magnitude of the gravitational force on

    P is $\dfrac{mgR^2}{x^2}$ **(3)**

Figure 1

A particle is fired vertically upwards with speed $2U$ from a point which is at a height $\dfrac{1}{2}R$ above
the surface of the Earth.

    b Show that the greatest height achieved above the Earth's surface is $\dfrac{(gR + 6U^2)R}{2(gR - 3U^2)}$ **(7)**

4   A particle P of mass m is attached to one end of a light elastic string of natural length a
    and modulus of elasticity $2mg$. The other end of the string is attached to a fixed point A.
    The string is initially resting in equilibrium with P hanging vertically below A. The particle

is then pulled by a horizontal force of magnitude $\frac{3}{4}mg$ so that the particle is now resting in equilibrium with the string making an angle θ with the downward vertical. The force is acting in the same vertical plane as the string.

Find the extension in the string. **(6)**

5 Figure 2 shows a particle P, of mass m, attached to the ends of two light inextensible strings AP and BP. The point A is a distance $10a$ vertically above B, and the two strings are taut and perpendicular to each other, with string AP longer than string BP. Particle P is performing horizontal circles with angular speed ω.

Given that the tension in AP is twice the tension in BP, find the value of ω. **(10)**

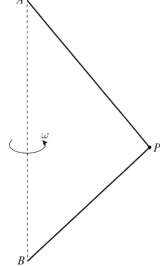

Figure 2

6 A light rod AB of length $\frac{3}{2}a$ has a particle P of mass $2m$ attached to B. The rod can rotate in a vertical plane about a smooth horizontal axis through A. The system is resting in equilibrium with P vertically below A. P is then projected horizontally with speed u.

  a Prove that in order for P to describe a complete vertical circle, then $u \geqslant \sqrt{\dfrac{15}{2}ga}$ **(8)**

  b Given that $u = \sqrt{5ga}$, find

    i the angle turned through when the speed of P is $\sqrt{ga}$

    ii the height of P above its starting point, giving your answer in the form ka, where k is to be given to 3 significant figures. **(5)**

7 A particle P of mass $2\,\text{kg}$ is attached to one end of a light elastic string, of natural length $1.2\,\text{m}$ and modulus of elasticity $49\,\text{N}$. The other end of the string is attached to a fixed point A on a ceiling. The particle rests in equilibrium at the point E, vertically below the point A with the string taut.

  a Find the extension in the string. **(4)**

The particle is projected downwards from its equilibrium position with speed $\frac{2}{35}\,\text{ms}^{-1}$.

  b Show that P moves with simple harmonic motion and determine the period of the motion in the form $\frac{a}{b}\pi\sqrt{\dfrac{3}{5}}$, where a and b are integers to be determined. **(6)**

  c How far does P travel before first coming to instantaneous rest. **(3)**

  d Find the time taken for the particle to reach $0.1\,\text{m}$ above the level of E for the first time. **(4)**

TOTAL FOR PAPER: 75 MARKS

GLOSSARY

acceleration positive rate of change of velocity

adjoining next to something or joined to it

amplitude the maximum extent of a vibration or **oscillation**, measured from the position of **equilibrium**

angular the **velocity** of a body rotating about a fixed point

arbitrary undetermined, not given a specific value

asymptote a line that a curve approaches but never quite reaches

attraction (force of) the force between two objects that draws them together

banked a *banked* turn is a turn in which the vehicle travels with one side higher than the other

bead a **particle** with a hole in it

coefficient of friction a measure of how rough a surface is, usually denoted μ

collinear points on the same straight line

composite made up of several different parts, or **elements**

compress to force an object into less space

compression the **force** acting on an object that is being squashed

conical pendulum a **pendulum** that performs **horizontal** circles about a centre that is **vertically** below a fixed point

conservation of energy law that states that the total **energy** of a system that remains constant throughout the motion

constant a **variable** that does not change

deceleration negative rate of change in **velocity**

decreasing reducing in value

density the measurement of **mass** per unit volume

descend to move downwards

directly proportional two variables that increase together, whilst keeping a constant ratio

displacement distance in a particular direction

elastic the ability to stretch

elastic energy the **energy** stored in an **elastic** body that has been stretched or compressed

elastic limit the length to which a spring or string can be stretched before it becomes deformed

elemental a smaller part of a larger object

energy the capacity to do work

equilibrium having zero **resultant force**

exert the use of force to affect an object

extension the extra length created when an **elastic** object is **stretched** beyond its natural length

finite with a limit; having a fixed size

force a push or a pull

friction resistance due to how rough the surface is on which a body is moving

frustum a cone with a smaller cone removed from the top

gravitation(al) a **force** which causes objects to be attracted to each other due to their mass

hemisphere half a sphere

hollow with an empty space inside

horizontal at right angles to the vertical

illustrate to make something clearer using pictures, diagrams, examples, etc.

inclined sloped, at an angle

increasing rising in value

inextensible doesn't **stretch**

infinite without limit

initially at the start

instantaneously happening immediately

interval the period of time between two actions or events

inversely proportional one **variable** that decreases as another increases

joule the unit of **energy**

kinetic energy the **energy** a body possesses as a result of its motion

lamina a two-dimensional object whose thickness can be ignored

limiting equilibrium the point at which **equilibirium** is about the broken

magnitude size

mass the measure of how much matter is in an object

maximum largest possible value

model an equation or set of equations used to describe an observed change

modulus of elasticity a value that measures the **resistance** of an object to being **stretched** or **compressed**

moment measures the turning effect of a **force** on a rigid body

natural length the length of an **elastic** object before any forces are applied to it

negligible too small to be significant

origin a point used as a reference

oscillation regular variation in position about a central point

parallel two lines that are the same distance apart at every point

particle an object with **negligible** dimensions

pendulum a weight **suspended** from a fixed point that has **periodic** motion

period the time it takes to complete one **oscillation**

perpendicular at right angles to something

plane a flat surface

potential energy the **energy** a body possesses as a result of its position

project [verb] to launch or fire an object in a particular direction

projectile an object that rises, then falls under the influence of gravity

proportional two variables that change together, whilst keeping a constant ratio

radian a unit of angle

refinement the process of improving a mathematical model

released set in motion

resistance opposition

resolve to split a **vector** into its components

resultant force the total of all forces applied to a single body, also called the net force

revolution a full rotation about a fixed point

rotate turn about a fixed point

simple harmonic motion motion that is periodic (i.e. not **constant**) and without loss of energy

slack not **stretched** or pulled tight

slope a surface with one end higher than another

smooth a without friction

stretch to extend an **elastic** object beyond its natural length

stretched an **elastic** object that is extended beyond its natural length

summation the addition of a sequence of numbers

suspend to hang from

symmetry an exact match between two halves

taut stretched; pulled tight

tension the pulling **force** passed via a string or other continuous object

thrust a pushing **force**, causing an object to move in a particular direction

topple to become unsteady and fall over

translation a function that moves an object a certain distance

uniform the same throughout

uppermost highest in position

variable a quantity that can change

vector a quantity with **magnitude** and direction

velocity the rate of change of **displacement**

vertex the point at which two or more edges meet, or the turning point of a curve

vertical straight up and down

work done energy transfer that occurs when an object is moved over a distance by an external **force**

ANSWERS

CHAPTER 1

Prior knowledge check

1 a $\dfrac{4}{3(2-3x)^2} + c$　**b** $\dfrac{4e^{3x}}{3} + c$　**c** $-\dfrac{\cos 5\pi x}{5\pi} + c$

2 $y = 3e^{\frac{x-1}{3(x+2)}}$

3 $\ln\dfrac{4}{3}$

Exercise 1A

1 $(16 - 12e^{-0.25t})\,\text{m s}^{-1}$

2 $v = t\sin t$, $x = \int t\sin t\,\mathrm{d}t$. Using integration by parts and the initial condition $x = 0$ at $t = 0$, we get $x = \sin t - t\cos t$

At $t = \dfrac{\pi}{2}$, $x = -\dfrac{\pi}{2}\cos\dfrac{\pi}{2} + \sin\dfrac{\pi}{2} = 0 + 1 = 1$.

3 $2\ln 3\,\text{m}$

4 $11.5\,\text{m}$ (3 s.f.)

5 a $6\sqrt{2}\,\text{m s}^{-2}$　**b** $x = \dfrac{4}{3}\sin 3t$　**c** $t = \dfrac{\pi}{3}$

6 $v = \dfrac{3}{2} - \dfrac{3}{2 + t^2}$

7 a

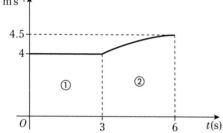

b $(27 - 3\ln 2)\,\text{m}$

8 a $4\,\text{m s}^{-1}$　　　　**b** $(\pi - 2\sqrt{2})\,\text{m}$

9 a $v = 40 - 20\,e^{0.2t}\,\text{m s}^{-1}$　**b** $(200\ln 2 - 100)\,\text{m}$

10 a $c = 80$, $d = 1$　　**b** $3200\ln\left(\dfrac{80 + t}{80}\right)$

11 a $t = \ln 2.5$, $\ln 3$　　**b** $\left(18 - \dfrac{15(\ln 3)^2}{2}\right)\text{m}$

12 a $t = 3$　　　　　　**b** $(12 + 3\ln 12)\,\text{m}$

13 a $t = 1$, $t = \dfrac{3}{2}$　**b** $a = 25\,\text{m s}^{-2}$　**c** $s = \dfrac{3917}{54}\,\text{m}$

d $x = t^3 - \dfrac{5}{2}t^2 + 2t = t\left(t^2 - \dfrac{5}{2}t + 2\right)$

For $t > 0$, when $t^2 - \dfrac{5}{2}t + 2 = 0$

The discriminant of the quadratic equation is

$\dfrac{25}{4} - (4 \times 2) < 0$

The equation has no real roots. Therefore, P never returns to the origin for any $t > 0$

14 a velocity (m s^{-1})

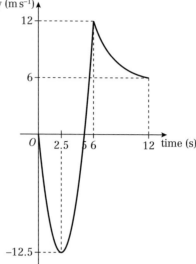

b $2.5 < t \leqslant 6$　　　**c** $\left(\dfrac{142}{3} + 72\ln(2)\right)\text{m}$

Challenge

$\dfrac{\mathrm{d}v}{\mathrm{d}t} = \dfrac{60}{kt^2}$

Integrating gives:

$v = -\dfrac{60}{kt} + c$

Solving for k and c using $v = 0$ when $t = 2$, and $v = 9$ when $t = 5$ gives $v = 15 - \dfrac{30}{t}$. As $t \geqslant 2$, $\dfrac{30}{t} > 0$ so $v < 15$ and the speed of the car never reaches $15\,\text{m s}^{-1}$

Exercise 1B

1 $v^2 = \dfrac{x^2}{2} + 4x + 25$

2 $v = \pm\sqrt{(80 - 4x^2)}$

3 $\dfrac{1}{5}$

4 $16\,\text{m}$

5 a $\dfrac{6}{125}$

b $\pm 4\sqrt{14}\,\text{m s}^{-1}$ as the particle will pass through this position in both directions.

6 4

7 a $v^2 = 52 - 36\cos\dfrac{x}{3}$

b $2\sqrt{22}\,\text{m s}^{-1}$ ($\approx 9.38\,\text{m s}^{-1}$)

8 $4.72\,\text{m s}^{-1}$ (3 s.f.), in the direction of x increasing.

9 a $1.95\,\text{m s}^{-1}$ (3 s.f.)　　**b** 26.8 (3 s.f.)

10 a $v = x + \dfrac{2}{x}$　　　　**b** $2\sqrt{2}\,\text{m s}^{-1}$

11 $10\,\text{m}$

12 a $v = 3x^{\frac{2}{3}}$　　　　　**b** $x = (t + 2)^3$

Challenge

$v^2 = \dfrac{1}{5}\left(25x - \dfrac{x^2}{2}\right) + \dfrac{163}{2}$

Chapter review 1

1 a $8(e^{0.5t} - 1)$ **b** $6\,\mathrm{m\,s^{-2}}$
2 a $v = 18 - 10e^{-t^2}$ **b** $18\,\mathrm{m\,s^{-1}}$
3 6
4 a $v = 10 - \dfrac{50}{2t + 5}$
 b $(100 - 25\ln 5)\,\mathrm{m} \approx 59.8\,\mathrm{m}$
5 a $\dfrac{\pi}{2}\,\mathrm{m\,s^{-1}}$
 b From part **a**, $v = \dfrac{1}{4}\sin 2t + \dfrac{1}{2}t$
 $x = \int v\,\mathrm{d}t = -\dfrac{1}{8}\cos 2t + \dfrac{1}{4}t^2 + c$
 Using inital conditions gives $c = \dfrac{1}{8}$, hence
 $x = -\dfrac{1}{8}\cos 2t + \dfrac{1}{4}t^2 + \dfrac{1}{8}$
 Substituting $t = \dfrac{\pi}{4}$ gives:
 $x = -\dfrac{1}{8}\cos\dfrac{\pi}{2} + \dfrac{1}{4}\left(\dfrac{\pi}{4}\right)^2 + \dfrac{1}{8} = \dfrac{1}{64}(\pi^2 + 8)$
6 a $2.5\,\mathrm{m\,s^{-2}}$ in the direction of x increasing.
 b $8\,e^{-1}\,\mathrm{m\,s^{-2}}$ in the direction of x decreasing.
 c $\left(\dfrac{56}{3} - 8\,e^{-2}\right)\mathrm{m} \approx 17.6\,\mathrm{m}$ (3 s.f.)
7 a $v = 2t + \ln(t + 1)$ **b** $(2 + 3\ln 3)\,\mathrm{m}$
8 a $T = 3\sqrt{10}\,\mathrm{s}$
 b velocity

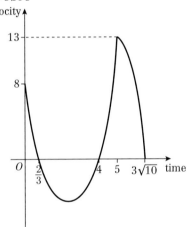

 c $\dfrac{7}{3} < t < 5$
 d $59.2\,\mathrm{m}$ (3 s.f.)
9 a $a = 6t - 20$ **b** $\dfrac{901}{27}\,\mathrm{m}$
 c The discriminant of the equation
 $x = t(t^2 - 10t + 32t) = 0$ is less than 0 for $t > 0$ and thus x can never be 0 for $t > 0$.
10 a $t = \dfrac{2}{3}$ or $t = 2$ **b** $\dfrac{248}{27}\,\mathrm{m}$
11 $\sqrt{\left(20 - \dfrac{12}{x}\right)}$
12 $x = 8$
13 $x = \dfrac{1}{5}$
14 a $v^2 = 16 + 6x - x^2$ **b** 5
15 a $v^2 = \dfrac{5x^3}{3} - \dfrac{x^4}{4} + \dfrac{2500}{3}$
 b $\dfrac{50\sqrt{3}}{3}\,\mathrm{m\,s^{-1}}$
16 a $6.04\,\mathrm{m\,s^{-1}}$ (3 s.f.) **b** 2.56 (3 s.f.)
17 a $7\,\mathrm{m\,s^{-1}}$ **b** $x = 7.56\,\mathrm{m}$ (3 s.f.)
18 a $v = 2x + 3$ **b** $x = \dfrac{3}{2}(e^{2t} - 1)$

Challenge
$11\,776\,\mathrm{km}$

CHAPTER 2
Prior knowledge check

1 $F = 2\sqrt{7}\,\mathrm{N}$, $\tan\theta = \dfrac{\sqrt{3}}{2}$ **2** $\mu = \dfrac{3}{g}$
3 $PQ = 2.55\,\mathrm{m}$ (3 s.f.)

Exercise 2A
1 a $3.4\,\mathrm{m}$ **b** $4\,\mathrm{m}$ **c** $3.75\,\mathrm{m}$
2 $4\,\mathrm{m}$ **3** $1.31\,\mathrm{m}$ (3 s.f.) **4** $\dfrac{mga}{\lambda}$
5 $l = \dfrac{m_1 a_2 - m_2 a_1}{m_1 - m_2}$
 $\lambda = g\dfrac{(m_1 a_2 - m_2 a_1)}{(a_1 - a_2)}$
6 $W = \dfrac{100}{3}\,\mathrm{N}$
7 a $\dfrac{11a}{4}$
 b If the spring is not light then in effect the mass would increase, the extension would increase and hence the distance of the particle below the ceiling would increase.
8 a $14.7\,\mathrm{N}$ **b** $2.7\,\mathrm{m}$ (2 s.f.)
9 $\dfrac{11l}{4}$
10 a $1\,\mathrm{m}$ **b** $0.58\,\mathrm{m}$ (2 s.f.) **c** $17\,\mathrm{N}$ (2 s.f.)
11 a $3.9\,\mathrm{N}$ (2 s.f.) **b** $0.96\,\mathrm{m}$ (2 s.f.)

Exercise 2B
1 a $6\,\mathrm{m\,s^{-2}}$ **b** $2\,\mathrm{m\,s^{-2}}$
2 $12.5\,\mathrm{m\,s^{-2}}$
3 $14.2\,\mathrm{m\,s^{-2}}$ upwards (3 s.f.)
4 $3.13\,\mathrm{m\,s^{-2}}$ downwards (3 s.f.)
5 a $4.62\,\mathrm{m\,s^{-2}}$ (3 s.f.)
 b Resultant force down plane $= T + g\sin\alpha - \mu R = ma$ so if μ increases acceleration would decrease.

Challenge
a Resolving vertically: $3g - 2T\cos 45 = \dfrac{3g}{2}$, gives $T = \dfrac{3\sqrt{2}\,g}{4}$
b $\lambda = 3\sqrt{2g}$

Exercise 2C
1 $1.07\,\mathrm{J}$ (3 s.f.) **2** $0.1\,\mathrm{J}$ **3** $1.125\,\mathrm{J}$
4 a $0.571\,\mathrm{J}$ (3 s.f.) **b** $1.14\,\mathrm{J}$ (3 s.f.) **c** $3.43\,\mathrm{J}$ (3 s.f.)
5 $23\,\mathrm{J}$ (2 s.f.)
6 $2mga$
7 a $T = mg\dfrac{\sqrt{41}}{4}\,\mathrm{N}$ **b** $\dfrac{41mga}{64}\,\mathrm{J}$

Exercise 2D
1 $V = \dfrac{1}{2}\sqrt{gl}$
2 $\dfrac{3a}{4} = d$
3 a Modulus is $mg\sqrt{3}$
 b Take into account the mass of the spring.
4 a $V = 2\,\mathrm{m\,s^{-1}}$ **b** $0.80\,\mathrm{m}$ (2 s.f.)
5 $U = \sqrt{\dfrac{3ag}{2}}$

6 a 160N (2 s.f.) **b** 2.9 m s⁻¹ (2 s.f.)
7 a Falls 4 m **b** 6.6 m s⁻¹ (2 s.f.)
8 0.11 (2 s.f.)

Challenge

When extension is $\frac{l}{10}$ E.P.E. $= \frac{Mgl}{20}$

When extension is doubled, E.P.E. $= \frac{4Mgl}{20}$

Work done is difference $= \frac{3Mgl}{20}$

Chapter review 2

1 a If $\cos\theta = \frac{4}{5}$ then $\sin\theta = \frac{3}{5}$

Resolving vertically: $2T\cos\theta = mg$, gives $T = \frac{5mg}{8}$

Using trigonometry: $\sin\theta = \frac{3}{5} = \frac{a}{a+x}$ gives $x = \frac{2a}{3}$

Substituting into $T = \frac{\lambda x}{a} = \frac{5mg}{8}$ gives $\lambda = \frac{15mg}{16}$

so $\cos\theta = \frac{4}{5}$

b $\frac{11mga}{12}$

2 $3a$

3 a $\lambda = 30$ N **b** $v = 2.19$ m s⁻¹ (3 s.f.)

4 $l = \frac{5\lambda a}{(5\lambda + 3mg)}$

5 a $V = \sqrt{\frac{13ag}{20}}$ **b** $d = \frac{13a}{50}$

6 a $\mu = \frac{2}{3}$ **b** $V = \sqrt{\frac{2gl}{3}}$ **c** $\frac{3l}{2}$

7 a extension of AP (x_1) $= 0.2\cos\theta - 0.15$;
extension of BP (x_2) $= 0.2\sin\theta - 0.05$;

$\Rightarrow$ ratio is $\frac{0.2\cos\theta - 0.15}{0.2\sin\theta - 0.05} = \frac{4\cos\theta - 3}{4\sin\theta - 1}$

b $T_2 = 5g\cos\theta$, $T_1 = 5g\sin\theta \Rightarrow \frac{T_2}{T_1} = \frac{\cos\theta}{\sin\theta}$

$\frac{\lambda x_2}{0.05} \times \frac{0.15}{\lambda x_1} = \frac{\cos\theta}{\sin\theta} \Rightarrow \frac{x_1}{x_2} = \frac{3\sin\theta}{\cos\theta}$

Using answer to part **a**, $\frac{4\cos\theta - 3}{4\sin\theta - 1} = \frac{3\sin\theta}{\cos\theta}$

Rearrange to arrive at required solution.

8 a $\theta = \tan^{-1}\left(\frac{1}{3}\right)$ **b** 2.1 m (2 s.f.) **c** 9.3 N (2 s.f.)

9 a When AP is vertical, $x = 4a$
K.E. gain + E.P.E. gain = P.E. loss

$\Rightarrow \frac{1}{2}mv^2 + \frac{mg}{4}\frac{x^2}{2a} = mg4a$

$\Rightarrow v = 2\sqrt{ga}$

b $T = mg$

Challenge

a If x is maximum distance, using conservation of energy
$mgx = \frac{\lambda(x-l)^2}{2l}$

Expanding and rearranging to form a quadratic in x
$\lambda x^2 - x(2l\lambda + 2mgl) + l^2\lambda = 0$

Use quadratic formula to arrive at required solution.

b i For a greater maximum descent the model could
include an initial velocity i.e. the person could jump
rather than fall.

ii For a smaller maximum descent air resistance
could be incorporated in to the model.

CHAPTER 3

Prior knowledge check

1 a 11170 N **b** 675 kJ
2 a 33.9 m s⁻¹ **b** 1.68 m
3 $13\frac{8}{9}$ N

Exercise 3A

1 a 9.09 m s⁻¹ (3 s.f.) **b** 1.41 m s⁻¹ (3 s.f.)
c P first comes to rest when $t = \pi$
d 14.2 m (3 s.f.) **e** $OP = 20$ m
2 a 10
b The van moves 10.6 m in the first 4 seconds (3 s.f.)
3 a Maximum speed occurs when acceleration is zero,
i.e. when force is zero. $\Rightarrow \frac{1}{6}(15 - x) = 0 \Rightarrow x = 15$
b 6.85 m s⁻¹
4 a 6.79 m s⁻¹ (3 s.f.) **b** 8.23 m s⁻¹ (3 s.f.)
c 8.10 N
5 $x = 0.677$ (3 s.f.)
6 a $0.25\frac{dv}{dt} = -\frac{8}{(t+1)^2} \Rightarrow \int 0.25\,dv = -\int\frac{8}{(t+1)^2}\,dt$

$\Rightarrow 0.25v = \frac{8}{(t+1)} + c \Rightarrow v = \frac{32}{(t+1)} + d$ where $d = 4c$

When $t = 0$, $v = 10$: $10 = \frac{32}{1} + d \Rightarrow d = -22$

$\Rightarrow v = 2\left(\frac{16}{(t+1)} - 11\right)$

b $x = 32\ln6 - 132$
7 $k = 66$

Challenge

a Work done $= \int_a^b 3x^2 - x^{\frac{1}{3}}\,dx = \left[x^3 - \frac{3}{4}x^{\frac{4}{3}}\right]_a^b$

$= b^3 - \frac{3}{4}b^{\frac{4}{3}} - a^3 + \frac{3}{4}a^{\frac{4}{3}}$

Hence work done is independent of the initial velocity.
b 208 J (3 s.f.)

Exercise 3B

1 $F = \frac{k}{d^2}$, where d = distance from centre

distance $(x - R)$ above surface
$\Rightarrow$ distance x from centre $\Rightarrow F = \frac{k}{x^2}$

On surface $F = mg$, $x = R \Rightarrow mg = \frac{k}{x^2} \Rightarrow k = mgR^2$

$\therefore$ Magnitude of the gravitational force is $\frac{mgR^2}{x^2}$

2 For a particle of mass m, distance x from the centre of
the earth.

$F = ma \Rightarrow \frac{k}{x^2} = mA$

On the surface of the earth, $x = R$, $A = g$

$\Rightarrow mg = \frac{k}{R^2} \Rightarrow k = mgR^2 \Rightarrow mA = \frac{mgR^2}{x^2} \Rightarrow A = \frac{gR^2}{x^2}$

3 $\sqrt{2gR}$

4 $\sqrt{\dfrac{U^2X + U^2R - 2gRX}{X + R}}$

5 $\sqrt{\dfrac{7gR}{5}}$

6 $2\sqrt{\dfrac{gR}{3}}$

Online Worked solutions are available in SolutionBank.

7 a $F = \dfrac{k}{x^2}$ when $x = R$, $F = mg$ so $mg = \dfrac{k}{R^2} \Rightarrow k = mgR^2$

gravitational force on $S = \dfrac{mgR^2}{x^2}$

b speed $= \sqrt{\dfrac{7gR}{2}}$

Challenge

a 5.98×10^{24} kg **b** 5500 kg m^{-3}

Exercise 3C

1 a $\dfrac{\pi}{2}$ s **b** 1.83 m s^{-1} (3 s.f.)

2 a 1 m **b** 1.36 m s^{-1} (3 s.f.)

3 a 5 m **b** π s

4 $\dfrac{4}{3}$ m s^{-1}

5 17.9 m s^{-1} (3 s.f.)

6 a 1.26 m (3 s.f.) **b** $x = 1.26 \sin 4t$

7 a 0.133 m (3 s.f.) **b** 0.0141 m (3 s.f.)

8 a 1.37 m (3 s.f.) **b** 0.684 s (3 s.f.)

9 a 1.00 m s^{-1} (3 s.f.) **b** 0.922 m s^{-1} (3 s.f.)

10 9.25 J (3 s.f.)

11 a 1.26 m s^{-1} (3 s.f.) **b** $\dfrac{2}{3}$ s to fall 0.6 m

12 0.0738 s (3 s.f.)

13 a $x = 4 \sin 2t \Rightarrow \dot{x} = 8 \cos 2t \Rightarrow \ddot{x} = -16 \sin 2t \Rightarrow \ddot{x} = 4x$
∴ S.H.M.

b $4, \dfrac{2\pi}{2} = \pi$ s **c** 8 m s^{-1} **d** $\dfrac{\pi}{6}$ **e** $\dfrac{\pi}{12}$

14 a $x = 3 \sin\left(4t + \dfrac{1}{2}\right) \Rightarrow \dot{x} = 12 \cos\left(4t + \dfrac{1}{2}\right)$

$\Rightarrow \ddot{x} = -48 \sin\left(4t + \dfrac{1}{2}\right) \Rightarrow \ddot{x} = 16x$ ∴ S.H.M.

b Amplitude $= 3$ m, Period $= \dfrac{\pi}{2}$ s

c $x = 1.44$ (3 s.f.) **d** 0.660 (3 s.f.)

15 a 11.51 a.m. (nearest minute)

b 8.39 p.m. (nearest minute)

16 P takes 0.823 s to travel directly from B to A (3 s.f.)

Challenge

$\ddot{x} = -\omega^2 x$ $v^2 = \omega^2(a^2 - x^2)$

$v_1^2 = \omega^2(a^2 - x_1^2)$ (1)

$v_2^2 = \omega^2(a^2 - x_2^2)$ (2)

(2) − (1): $v_2^2 - v_1^2 = \omega^2(a^2 - x_2^2) - \omega^2(a^2 - x_1^2)$

$v_2^2 - v_1^2 = \omega^2(a^2 - x_2^2 - a^2 + x_1^2)$

Rearranging gives $\omega^2 = \dfrac{v_2^2 - v_1^2}{x_1^2 - x_2^2}$ so $\omega = \left(\dfrac{v_2^2 - v_1^2}{x_1^2 - x_2^2}\right)^{\frac{1}{2}}$

$T = \dfrac{2\pi}{\omega} = 2\pi\left(\dfrac{x_1^2 - x_2^2}{v_2^2 - v_1^2}\right)^{\frac{1}{2}}$

Exercise 3D

1 a $F = ma \Rightarrow -T = 0.5\ddot{x}$

Hooke's law: $T = \dfrac{\lambda x}{l} = \dfrac{60x}{0.6} = 100x$

$\Rightarrow -100x = 0.5\ddot{x} \Rightarrow \ddot{x} = -200x$ ∴ S.H.M.

b $\dfrac{\pi}{10}\sqrt{2}$ s 0.3 m **c** 4.24 m s^{-1} (3 s.f.)

2 a $F = ma \Rightarrow -T = 0.8\ddot{x}$

Hooke's law: $T = \dfrac{\lambda x}{l} = \dfrac{20x}{1.6} = \dfrac{25x}{2}$

$\Rightarrow -\dfrac{25x}{2} = 0.8\ddot{x} \Rightarrow \ddot{x} = -\dfrac{125x}{8}$ ∴ S.H.M.

b 3.21 s (3 s.f.)

3 a $F = ma \Rightarrow -T = 0.4\ddot{x}$

Hooke's law: $T = \dfrac{\lambda x}{l} = \dfrac{24x}{1.2} = 20x$

$\Rightarrow -20x = 0.4\ddot{x} \Rightarrow \ddot{x} = -50x$ ∴ S.H.M.

b 0.489 s (3 s.f.) **c** 1.84 m (3 s.f.)

4 a $F = ma \Rightarrow -T = 0.8\ddot{x}$

Hooke's law: $T = \dfrac{\lambda x}{l} = \dfrac{80x}{1.2}$

$\Rightarrow -\dfrac{80x}{1.2} = 0.8\ddot{x} \Rightarrow \ddot{x} = -\dfrac{100x}{1.2}$ ∴ S.H.M.

b period $= \dfrac{6\pi}{5\sqrt{30}}$ seconds, amplitude $= 0.6$ metres

c $v_{max} = \sqrt{30}$ m s^{-1}

5 a $x = 0.5 \sin 10t$ **b** 50 m s^{-2}

6 a 0.5 m **b** 2.11 m s^{-1} (3 s.f.)

c i 1.49 s **ii** 0.3 m

7 a $0.351s$ (3 s.f.) **b** 2.56 J

8 a 2.45 m s^{-1} (3 s.f.) **b** 1.25 (3 s.f.)

9 a $F = ma \Rightarrow T_B - T_A = 0.4\ddot{x}$

Hooke's law: $T = \dfrac{\lambda x}{l}$

AP: extension $= 0.8 + x$

$\therefore T_A = \dfrac{12(0.8 + x)}{1.2} = 10(0.8 + x)$

BP: extension $= 0.8 - x$

$\therefore T_B = \dfrac{12(0.8 - x)}{1.2} = 10(0.8 - x)$

$\therefore 10(0.8 - x) - 10(0.8 + x) = 0.4\ddot{x}$
$-20x = 0.4\ddot{x} \Rightarrow \ddot{x} = -50x$ ∴ S.H.M.

b 3.6 J

10 a $F = ma \Rightarrow T_B - T_A = m\ddot{x}$

Hooke's law: $T = \dfrac{\lambda x}{l}$

AP: extension $= 1.5l + x \Rightarrow T_A = \dfrac{3mg(1.5l + x)}{l}$

PB: extension $= 1.5l - x \Rightarrow T_B = \dfrac{3mg(1.5l - x)}{l}$

$\Rightarrow \dfrac{3mg(1.5l + x)}{l} - \dfrac{3mg(1.5l - x)}{l} = m\ddot{x}$

$-\dfrac{6mgx}{l} = m\ddot{x} \Rightarrow \ddot{x} = -\dfrac{6g}{l}x$ ∴ S.H.M.

b $2\pi\sqrt{\dfrac{l}{6g}}$ **c** $1.5l$

d $\sqrt{12gl}$ (or $2\sqrt{3gl}$)

11 a $F = ma \Rightarrow T_B - T_A = 0.5\ddot{x}$

Hooke's law: $T = \dfrac{\lambda x}{l}$

AP: extension $= 1 + x \therefore T_A = \dfrac{15(1 + x)}{1} = 15(1 + x)$

BP: extension $= 1.5 - x$

$\therefore T_B = \dfrac{15(1.5 - x)}{1.5} = 10(1.5 - x)$

$\therefore 10(1.5 - x) - 15(1 + x) = 0.5\ddot{x}$
$-25x = 0.5\ddot{x} \Rightarrow \ddot{x} = -50x$ ∴ S.H.M.

period $= \dfrac{2\pi}{\omega} = \dfrac{2\pi}{\sqrt{50}} = \dfrac{\pi}{5}\sqrt{2}$

b 1 m

Exercise 3E

1 a 1.64 m (3 s.f.)

b $F = ma \Rightarrow 0.75g - T = 0.75\ddot{x}$

Hooke's law: $T = \dfrac{\lambda x}{l} = \dfrac{80(x + e)}{1.5}$

$\therefore 0.75g - \dfrac{80(x + e)}{1.5} = 0.75\ddot{x}$

from part **a**, $0.75g = \dfrac{80e}{1.5}$

$\therefore 0.75\ddot{x} = -\dfrac{80}{1.5}x \Rightarrow \ddot{x} = -\dfrac{80x}{1.5 \times 0.75}$ $\therefore$ S.H.M.

c 0.745 s (3 s.f.)

d 0.296 m (3 s.f.)

2 a 0.049 m (3 s.f.)

b 0.444 s (3 s.f.)

c 2.83 m s^{-1} (3 s.f.)

3 a 1.5 m s^{-1}

b In equilibrium:

R($\uparrow$): $T = 2mg$

Hooke's Law: $T = \dfrac{\lambda x}{l} = \dfrac{\lambda e}{1.5}$ $\therefore \dfrac{\lambda e}{1.5} = 2g$

When oscillating: $F = ma \Rightarrow 2g - T = 2\ddot{x}$

Hooke's Law: $T = \dfrac{\lambda(e + x)}{1.5}$

$\therefore 2g - \dfrac{\lambda(e + x)}{1.5} = 2\ddot{x}$

$\Rightarrow \dfrac{\lambda e}{1.5} - \dfrac{\lambda(e + x)}{1.5} = 2\ddot{x} \Rightarrow -\dfrac{\lambda x}{1.5} = 2\ddot{x}$

$\therefore \ddot{x} = -\dfrac{\lambda}{3}x$ as $\lambda > 0$, this is S.H.M.

c 48 **d** 0.375 m

4 a 0.138 s (3 s.f.) **b** 1.61 m s^{-1}

5 a 31.4 N (3 s.f.)

b $F = ma \Rightarrow 0.4g - T = 0.4\ddot{x}$

Hooke's law: $T = \dfrac{\lambda x}{l} = \dfrac{31.36(x + 0.05)}{0.4}$

$0.4g - \dfrac{31.36(x + 0.05)}{0.4} = 0.4\ddot{x}$

$\ddot{x} = -\dfrac{31.36}{0.4^2}x$ $\therefore$ S.H.M.

c 0.449 s, 0.07 m **d** 0.98 m s^{-1}

e 0.156 s to rise 11 cm (3 s.f.)

6 a 0.416 s (3 s.f.) **b** 0.0574 s (3 s.f.)

7 a 0.221 m **b** 0.517 s (3 s.f.)

8 a 1.69 m (3 s.f.)

b $T_A = \dfrac{15(e + x)}{0.6}$, $T_B = \dfrac{15(1.6 - (e + x))}{0.6}$

$F = ma$

$1.5g + \dfrac{15(1.6 - (e + x))}{0.6} - \dfrac{15(e + x)}{0.6} = 1.5\ddot{x}$

$1.5g + 40 - 25(e + x) - 25(e + x) = 1.5\ddot{x}$

$1.5g + 40 - 50e - 50x = 1.5\ddot{x}$

From part **a**, $50e = 1.5g + 40$

$\therefore 1.5\ddot{x} = -50x \Rightarrow \ddot{x} = -\dfrac{100}{3}x$ $\therefore$ S.H.M.

c 0.398 (3 s.f.)

9 a 12.5 m s^{-1} (3 s.f.) **b** 10.4 m (3 s.f.)

c 1.56 s (3 s.f.)

Challenge

Particle P: $T = 2\pi\sqrt{\dfrac{ml}{\lambda}} = 2\pi\sqrt{\dfrac{ml}{5mg}} = 2\pi\sqrt{\dfrac{l}{5g}}$

Particle Q: $3T = 2\pi\sqrt{\dfrac{l(m + km)}{5mg}} = 2\pi\sqrt{\dfrac{l(1 + k)}{5g}}$

$6\pi\sqrt{\dfrac{l}{5g}} = 2\pi\sqrt{\dfrac{l(1 + k)}{5g}}$ so $3\sqrt{l} = 2\sqrt{l} \times \sqrt{1 + k}$

squaring both sides $9 = 4 + 4k$ so $k = \dfrac{5}{4}$

Chapter review 3

1 a 108

b 11.8 m (3 s.f.)

2 a $\dot{x} = \dfrac{-20}{3}(3t + 4)^{\frac{1}{2}} + \dfrac{76}{3}$

b 18.7 m from O (3 s.f.)

3 a $F = \dfrac{k}{x^2}$ when $x = R$, $F = mg$

$\therefore \dfrac{k}{x^2} = mg$, $k = mgR^2$

b $\sqrt{\dfrac{8Rg}{5}}$ or $2\sqrt{\dfrac{2Rg}{5}}$

4 a $\dfrac{4\pi}{50}$ (or 0.251 (3 s.f.))

b 0.203 m s^{-1} (3 s.f.)

c 0.318 m (3 s.f.)

5 a $x = 2.5$

b $v^2 = 25x - 5x^2 + 32.75$

6 a $x = 3\sin\left(\dfrac{\pi}{4}t\right) \Rightarrow \dot{x} = \dfrac{3\pi}{4}\cos\left(\dfrac{\pi}{4}t\right)$

$\Rightarrow \ddot{x} = -3\left(\dfrac{\pi}{4}\right)^2\sin\left(\dfrac{\pi}{4}t\right)$

$\Rightarrow \ddot{x} = -\left(\dfrac{\pi}{4}\right)^2 x$ $\therefore$ S.H.M.

b Amplitude = 3 m

Period = 8 s

c $\dfrac{3\pi}{4}$ m s^{-1} (or 2.36 m s^{-1} (3 s.f.))

d 0.405 s (3 s.f.)

7 a 1.54 s

b 0.116 m

c 1.03 s

8 a $F = ma \Rightarrow -T = 0.6\ddot{x}$

Hooke's law: $T = \dfrac{\lambda x}{l} = \dfrac{25x}{2.5} = 10x$

$\Rightarrow -10x = 0.6\ddot{x} \Rightarrow \ddot{x} = -\dfrac{10}{0.6}x$ $\therefore$ S.H.M.

b Period is 1.54 s (3 s.f.)

Amplitude is $(4 - 2.5)$ m = 1.5 m

c P takes 0.468 s to move 2 m from B (3 s.f.)

9 a $F = ma \Rightarrow T_B - T_A = 0.4\ddot{x}$

Hooke's law: $T = \dfrac{\lambda x}{l}$

$T_A = \dfrac{2.5(0.4 + x)}{0.6}$, $T_B = \dfrac{2.5(0.4 - x)}{0.6}$

$\therefore \dfrac{2.5(0.4 - x)}{0.6} - \dfrac{2.5(0.4 + x)}{0.6} = 0.4\ddot{x}$

$\Rightarrow -5x = 0.4\ddot{x} \Rightarrow \ddot{x} = -\dfrac{5}{0.6 \times 0.4}x$ $\therefore$ S.H.M.

b 1.38 s (3 s.f.)

c 0.229 s to reach D (3 s.f.)

10 a Spring AP extension: 2.4 m

Spring PB extension: 1.6 m

b $F = ma$

$T_B - T_A = 0.4\ddot{x}$

$T_A = \dfrac{20(2.4 + x)}{5}$

$T_B = \dfrac{18(1.6 - x)}{3}$

$6(1.6 - x) - 4(2.4 + x) = 0.4x$

$-10x = 0.4\ddot{x}$

$\ddot{x} = -25x$

$\therefore$ S.H.M.

c 4 m s^{-1}

Online Worked solutions are available in SolutionBank.

11 a $\lambda = 3g$

b $0.5g - T = 0.5\ddot{x}$, $T = \dfrac{\lambda x}{l} = \dfrac{3g(0.2 + x)}{1.2}$

$\Rightarrow 5g - \dfrac{3g(0.2 + x)}{1.2} = 0.5\ddot{x} \therefore \ddot{x} = \dfrac{-3gx}{0.5 \times 1.2} = -5gx$

$\therefore$ S.H.M.

c $0.898\,\text{s}$ (3 s.f.)

d $2.01\,\text{m s}^{-1}$ (3 s.f.)

e $0.406\,\text{m}$ (3 s.f.)

12 a $F = ma \Rightarrow T_B - T_A = m\ddot{x}$

Hooke's law: $T = \dfrac{\lambda x}{l}$

$T_A = \dfrac{5mg(l - x)}{2l}$, $T_B = \dfrac{5mg(l + x)}{2l}$

$\therefore \dfrac{5mg(l - x)}{2l} - \dfrac{5mg(l + x)}{2l} = m\ddot{x}$

$\Rightarrow -\dfrac{5mgx}{l} = m\ddot{x} \Rightarrow \ddot{x} = -\dfrac{5g}{l}x \therefore$ S.H.M.

b $2\pi\sqrt{\dfrac{l}{5g}}$ **c** $\dfrac{1}{4}\sqrt{5gl}$

Challenge

$ma = -\dfrac{mMG}{(R + x)^2}$ so $a = -\dfrac{MG}{(R + x)^2}$

a $v\dfrac{dv}{dx} = -\dfrac{MG}{(R + x)^2}$ maximum height, H, is reached

when $v = 0$

$\displaystyle\int_u^0 v\, dv = -\int_0^H \dfrac{MG}{(R + x)^2}\, dx \Rightarrow \left[\dfrac{v^2}{2}\right]_u^0 = \left[\dfrac{MG}{(R + x)}\right]_0^H$

$-\dfrac{u^2}{2} = \dfrac{MG}{(R + H)} - \dfrac{MG}{R} = MG\left(\dfrac{1}{R + H} - \dfrac{1}{R}\right) = MG\left(\dfrac{R - R - H}{R(R + H)}\right)$

$-\dfrac{u^2}{2} = \dfrac{MG}{(R + H)} - \dfrac{MG}{R} = MG\left(\dfrac{1}{R + H} - \dfrac{1}{R}\right) = MG\left(\dfrac{R - R - H}{R(R + H)}\right)$

$MG\left(\dfrac{H}{R(R + H)}\right) = \dfrac{u^2}{2} \Rightarrow 2MGH = u^2(R^2 + RH)$

$2MGH - RHu^2 = u^2R^2 \quad H(2MG - Ru^2) = u^2R^2$

$H = \dfrac{R^2 u^2}{2MG - Ru^2} = \dfrac{Ru^2}{\dfrac{2MG}{R} - u^2}$

b $H \to \infty$ as $u^2 \to \dfrac{2MG}{R}$

Escape velocity $u = \sqrt{\dfrac{2MG}{R}} = \sqrt{\dfrac{2 \times 5.98 \times 10^{24} \times 6.7 \times 10^{11}}{6.4 \times 10^6}}$

$= 1.12 \times 10^4\,\text{m s}^{-1}$

Review exercise 1

1 $a = \dfrac{dv}{dt} = e^{2t}$

$v = \int e^{2t}\, dt = \dfrac{1}{2}e^{2t} + A$

When $t = 0, v = 0$

$0 = \dfrac{1}{2} + A \Rightarrow A = -\dfrac{1}{2}$

$v = \dfrac{1}{2}(e^{2t} - 1)$

2 a $v = 13 - 3e^{\frac{1}{6}t}$ **b** $11.2\,\text{m s}^{-1}$ (3 s.f.)   **c** 13

3 a $v = 8 - 4\cos\dfrac{1}{2}t$ **b** $4(\pi - \sqrt{2})\,\text{m}$

4 a $a = \dfrac{dv}{dt} = 3(t + 4)^{-\frac{1}{2}}$

$v = -3\int (t + 4)^{-\frac{1}{2}}\, dt = A - 6(t + 4)^{\frac{1}{2}}$

When $t = 0, v = 18$

$18 = A - 6 \times 2 \Rightarrow A = 30$

$v = 30 - 6\sqrt{t + 4}$

b $162\,\text{m}$

5 a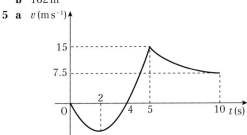

b $2 < t < 5$

c $\displaystyle\int_0^4 3t(t - 4)\, dt = [t^3 - 12t^2]_0^4 = -32$

$\displaystyle\int_4^5 3t(t - 4)\, dt = [t^3 - 12t^2]_4^5 = 7$

So distance travelled in the interval is $32 + 7 = 39\,\text{m}$

d 6.98 (3 s.f.)

6 $a = \dfrac{d}{dx}\left(\dfrac{1}{2}v^2\right) = 4x$

$\dfrac{1}{2}v^2 = \int 4x\, dx = 2x^2 + A$

$v^2 = 4x^2 + B$

At $x = 2, v = 4$

$16 = 16 + B \Rightarrow B = 0$

$v^2 = 4x^2$

7 $a = \dfrac{d}{dx}\left(\dfrac{1}{2}v^2\right) = 1 - 4x^{-2}$

$\dfrac{1}{2}v^2 = \int(1 - 4x^{-2})\, dx$

$= x + \dfrac{4}{x} + A$

$v^2 = 2x + \dfrac{8}{x} + B$, where $B = 2A$

At $x = 1, v = 3\sqrt{2}$

$18 = 2 + 8 + B \Rightarrow B = 8$

$v^2 = 2x + \dfrac{8}{x} + 8$

At $x = \dfrac{3}{2}$

$v^2 = 2 \times \dfrac{3}{2} + 8 \times \dfrac{2}{3} + 8 = \dfrac{49}{3}$

8 $8.76\,\text{m s}^{-1}$ (3 s.f.)

9 a $v^2 = 4k^2\left(1 - \dfrac{2}{x + 1}\right)$

b $v = 2k\sqrt{1 - \dfrac{2}{x + 1}}$

As x is positive, $\dfrac{1}{1 + x}$ is positive and $1 - \dfrac{1}{x + 1} < 1$

So $v < 2k$

10 a At the maximum value of v, $a = 0$

$a = \dfrac{1}{12}(30 - x) = 0$

$x = 30$

b $v^2 = 5x - \dfrac{x^2}{12} + 25$

11 a $0.9\,\text{m}$

b $a = p - qx$

At $x = 0, a = 20$

$20 = p - 0 \Rightarrow p = 20$

$a = 20 - qx$

At $x = 2, a = 12$

$12 = 10 - 2q \Rightarrow q = 4$

c $1\,\text{m}$

12 a $1.05\,\text{m}$ **b** $7.5\,\text{N}$

13 $AB = 2.55\,\text{m}$, $BC = 1.45\,\text{m}$

14 a The line of action of the weight must pass through C which is not above the centre of the rod.

b Let the tension in AC be T_1 newtons and the tension in BC be T_2 newtons.

R($\uparrow$): $T_1\cos\alpha = T_2\sin\alpha \Rightarrow T_1 = \dfrac{3}{4}T_2$

R($\rightarrow$): $T_1\sin\alpha + T_2\cos\alpha = 2mg$

$\dfrac{3}{4}T_2 \times \dfrac{3}{5} + T_2 \times \dfrac{4}{5} = 2mg \Rightarrow T_2 = \dfrac{8}{5}mg$

c $k = 8$

15 $a = 11\,\mathrm{m\,s^{-1}}$ (2 s.f.)

16 $\lambda = \dfrac{15}{16}mg$

17 a $\dfrac{5\sqrt{3}}{3}\,\mathrm{m\,s^{-1}} \approx 2.89\,\mathrm{m\,s^{-1}}$ (3 s.f.)

b 7.79 N (3 s.f.)

18 By work–energy principle:

E.P.E. = Work done $= \dfrac{\lambda x^2}{2l}$

From Hooke's law:

$T = \dfrac{\lambda x}{l} = mg \Rightarrow x = \dfrac{mgl}{\lambda}$

E.P.E. $= \dfrac{\lambda m^2 g^2 l^2}{2l\lambda^2} = \dfrac{m^2 g^2 l}{2\lambda}$

19 2.85 J (3 s.f.)

20 a 14 (2 s.f.) **b** 0.78 m (2 s.f.) **c** 6.1 J (2 s.f.)

21 a $0.2mga^2$ **b** $\mu = 0.6$

22 $\dfrac{3}{4}a$

23 $\dfrac{1}{2}mv^2 = \dfrac{\lambda x^2}{2l} \Rightarrow 2.5v^2 = \dfrac{75 \times 0.5^2}{2 \times 1} \Rightarrow v = \dfrac{\sqrt{15}}{2}$

24 a $5.5\,\mathrm{m\,s^{-1}}$ (2 s.f.) **b** $3.8\,\mathrm{m\,s^{-1}}$ (2 s.f.)

25 a 6 m **b** $14\,\mathrm{m\,s^{-1}}$ (2 s.f.)

26 a $v^2 = 20 - 16e^{-0.1x}$

b $x = 10\ln 4$

c For all x, $e^{-0.1x} > 0$
So $v^2 = 20 - 16e^{-0.1x} < 20$
Hence $v < \sqrt{20}$

27 a $v^2 = 2x^2 - x^3 + 144$ **b** $12\,\mathrm{m\,s^{-1}}$

28 a $F = ma$

$\dfrac{48000}{(t+2)^2} = 800a$

$a = 60(t+2)^{-2}$

$v = \int 60(t+2)^{-2}\,dt$

$= A - \dfrac{60}{t+2}$

When $t = 0$ $v = 0$

$0 = A - \dfrac{60}{2} \Rightarrow A = 30$

$v = 30 - \dfrac{60}{t+2}$

As $t \to \infty$, $\dfrac{60}{t+2} \to 0$, $v \to 30$

As t increases, the car approached a limiting speed of $30\,\mathrm{m\,s^{-2}}$

b $(180 - 120\ln 2)\,\mathrm{m}$

29 a 6 **b** 17 m

30 a As $F \propto \dfrac{1}{x^2}$, $F \propto -\dfrac{k}{x^2}$

At $x = R$, $F = -mg$

$-mg = -\dfrac{k}{R^2} \Rightarrow k = mgR^2$

$F = -\dfrac{mgR^2}{x^2}$

b $\sqrt{\dfrac{gR}{6}}$

31 a $F = ma$

$-\dfrac{k}{x^2} = ma$

$a = -\dfrac{k}{ma^2}$

At $x = R$, $a = -g$

$-g = -\dfrac{k}{mR^2}$

$k = mgR^2$

$-\dfrac{mgR^2}{x^2} = ma$

$a = v\dfrac{dv}{dx} = -\dfrac{gR^2}{x^2}$

b $X = \dfrac{2gR^2}{2gR - U^2}$

32 $\dfrac{2}{3}$

33 a 1.3 m **b** $2.6\,\mathrm{m\,s^{-1}}$

c $5.2\,\mathrm{m\,s^{-2}}$ **d** 0.79 s (2 d.p.)

34 a At C: $v^2 = \omega^2(a^2 - x^2)$

$0 = \omega^2(a^2 - 1.2^2) \Rightarrow a = 1.2$

At A: $v^2 = \omega^2(a^2 - x^2)$

$\left(\dfrac{3}{10}\sqrt{3}\right)^2 = \omega^2(1.2^2 - 0.6^2)$

$\omega^2 = \dfrac{27}{108} = \dfrac{1}{4} \Rightarrow \omega = \dfrac{1}{2}$

Checking $a = 1.2$ and $\omega = \dfrac{1}{2}$ at B

$v^2 = \omega^2(a^2 - x^2)$

$= \dfrac{1}{4}(1.2^2 - 0.8^2) = \dfrac{1}{5}$

$v = \dfrac{1}{5}\sqrt{5}$

b At O, $x = 0$:

$v^2 = \dfrac{1}{4}(1.2^2 - 0^2) = 0.36$

$v = 0.6\,\mathrm{m\,s^{-2}}$

c $0.15\,\mathrm{m\,s^{-2}}$ **d** 0.412 s (3 s.f.)

35 a Let the piston be modelled by the particle P.
Let O be the point where $AO = 0.6\,\mathrm{m}$
When P is at a general point in its motion.
Let $OP = x$ metres and the force of the spring on P be T newtons.
Hooke's Law:

$T = \dfrac{\lambda x}{l} = \dfrac{48x}{0.6} = 80x$

R($\rightarrow$): $F = ma$

$-T = 0.2\ddot{x}$

$-80x = 0.2\ddot{x}$

$\ddot{x} = -400x = -20^2 x$

Comparing with the standard formula for simple harmonic motion, this is simple harmonic motion with $\omega = 20$.

$T = \dfrac{2\pi}{20} = \dfrac{\pi}{10}$

b $6\,\mathrm{m\,s^{-1}}$ **c** $\dfrac{\pi}{15}\,\mathrm{s}$

36 a Let E be the point where $OE = 0.6\,\mathrm{m}$
When P is at a general point in its motion, let $EP = x$ metres and the force of the spring on P be T newtons.

Online Worked solutions are available in SolutionBank.

Hooke's Law:

$T = \dfrac{\lambda x}{l} = \dfrac{12x}{0.6} = 20x$

$R(\rightarrow)$: $F = ma$

$\qquad -T = 0.8\ddot{x}$

$\qquad -20x = 0.8\ddot{x}$

$\ddot{x} = -25x = -5^2 x$

Comparing with the standard formula for simple harmonic motion, this is simple harmonic motion with $\omega = 5$.

$T = \dfrac{2\pi}{\omega} = \dfrac{2\pi}{5}$

b $6.25\,\text{m s}^{-2}$ **c** $0.68\,\text{m s}^{-1}$ (2 s.f.)

d As it passes through C, P is moving away from O towards B.

37 a When P is at the point X, $AX = x$ metres and the force of the spring be T newtons.

Hooke's Law:

$T = \dfrac{\lambda x}{l} = \dfrac{21.6x}{2} = 10.8x$

$R(\rightarrow)$: $F = ma$

$\qquad -T = 0.3\ddot{x}$

$\qquad -10.8x = 0.3\ddot{x}$

$\ddot{x} = -36x = -6^2 x$

Comparing with the standard formula for simple harmonic motion, this is simple harmonic motion with $\omega = 6$.

$T = \dfrac{2\pi}{\omega} = \dfrac{2\pi}{6} = \dfrac{\pi}{3}$

b $9\,\text{m s}^{-1}$ **c** $\dfrac{\pi}{18}\,\text{s}$ **d** $1.16\,\text{m}$ (3 s.f.)

38 a $\dfrac{9a}{2}$

b When P is at a general point X, let $OX = x$.

At this point the extension of the string is $0.5a + x$.

Hooke's Law:

$T = \dfrac{\lambda e}{l} = \dfrac{8mg\left(\frac{1}{2}a + x\right)}{a} = mg + \dfrac{2mgx}{a}$ (1)

Newton's second law:

$R(\downarrow)$: $F = ma$

$mg - t = m\dfrac{d^2 x}{dt^2}$ (2)

Substitute (1) into (2):

$mg - \left(mg + \dfrac{2mgx}{a}\right) = m\dfrac{d^2 x}{dt^2}$

$\qquad -\dfrac{2mgx}{a} = m\dfrac{d^2 x}{dt^2}$

$\qquad \dfrac{d^2 x}{dt^2} = -\dfrac{2gx}{a}$

Comparing with the standard formula for simple harmonic motion, this is simple harmonic motion with $\omega = \sqrt{\dfrac{2g}{a}}$

$T = \dfrac{2\pi}{\omega} = 2\pi\sqrt{\dfrac{a}{2g}} = \pi\sqrt{\dfrac{2a}{g}}$

c $\dfrac{1}{2\sqrt{2}}a$

d As $a > \dfrac{1}{2}a$, the string will become slack during its motion. The subsequent motion of P will be partly under gravity, partly simple harmonic motion.

Challenge

1 $a = 8x\dfrac{dx}{dt}$

Using $a = v\dfrac{dv}{dx}$ and $v = \dfrac{dx}{dt}$

$v\dfrac{dv}{dx} = 8xv$

$\dfrac{dv}{dx} = 8x$

$\int dv = \int 8x\,dx$

$\qquad v = 4x^2 + c$

$t = 0, x = 0, v = -k: -k = c$

$\dfrac{dx}{dt} = 4x^2 - k$

Displacement has maximum when $\dfrac{dx}{dt} = 0$

$4x^2 - k = 0 \Rightarrow x = \dfrac{\sqrt{k}}{2}$

So maximum displacement is $\dfrac{1}{2}\sqrt{k}$

2 a Work done $= \int_0^x T\,ds$

$\qquad = \int_0^x \dfrac{\lambda s}{l}\,ds$

$\qquad = \left[\dfrac{\lambda s^2}{2l}\right]_0^x$

$\qquad = \dfrac{\lambda x^2}{2l}$

b Work done $=$ E.P.E. gain of string

$\qquad = \dfrac{\lambda}{2l}(b^2 - a^2)$

$\qquad = \dfrac{\lambda}{2l}(b + a)(b - a)$

$\qquad = \dfrac{1}{2}\left(\dfrac{\lambda b}{l} + \dfrac{\lambda a}{l}\right)(b - a)$

$\qquad = \dfrac{1}{2}(T_b + T_a)(b - a)$

$\qquad = $ mean of tensions $\times$ distance moved

CHAPTER 4
Prior knowledge check

1 a $21.9\,\text{N}$ (3 s.f.)   **b** $3.05\,\text{kg}$ (3 s.f.)

2 a $10.6\,\text{m}$ (3 s.f.)   **b** $58.6\,\text{J}$ (3 s.f.)

Exercise 4A

1 a $0.524\,\text{rad s}^{-1}$ (3 s.f.)   **b** $12.6\,\text{rad s}^{-1}$ (3 s.f.)

   c $38.2\,\text{rev min}^{-1}$ (3 s.f.)   **d** $1720\,\text{rev h}^{-1}$ (3 s.f.)

2 a $80\,\text{m s}^{-1}$ **b** $83.8\,\text{m s}^{-1}$ (3 s.f.)

3 a $8\,\text{rad s}^{-1}$ **b** $76.4\,\text{rev min}^{-1}$ (3 s.f.)

4 a $2\,\text{m s}^{-1}$ **b** $2.09\,\text{m s}^{-1}$ (3 s.f.)

5 a $44.9\,\text{s}$ (3 s.f.)   **b** $0.14\,\text{rad s}^{-1}$ (3 s.f.)

6 a $0.628\,\text{rad s}^{-1}$ (3 s.f.)   **b** $0.0754\,\text{m s}^{-1}$ (3 s.f.)

   c $0.0503\,\text{m s}^{-1}$ (3 s.f.)

7 a $0.279\,\text{rad s}^{-1}$ (3 s.f.)   **b** $39.8\,\text{m}$ (3 s.f.)

8 $3.14\,\text{m s}^{-1}$ (3 s.f.), $5.24\,\text{m s}^{-1}$ (3 s.f.)

9 a $0.242\,\text{rad s}^{-1}$ (3 s.f.)   **b** $0.362\,\text{m s}^{-1}$ (3 s.f.)

10 $0.056\,\text{rad s}^{-1}$ (3 s.f.)

11 a $0.000145\,\text{rad s}^{-1}$ (3 s.f.), $0.00175\,\text{rad s}^{-1}$ (3 s.f.)

   b $1.45 \times 10^{-5}\,\text{m s}^{-1}$ (3 s.f.), $2.62 \times 10^{-4}\,\text{m s}^{-1}$ (3 s.f.)

12 $62.8\,\text{m s}^{-1}$ (3 s.f.)

13 a $4.71\,\text{rad s}^{-1}$ (3 s.f.)   **b** $2.55\,\text{cm}$ (3 s.f.)

14 $29\,900\,\text{m s}^{-1}$ (3 s.f.)

15 $r > 5$

Challenge

$\frac{\pi}{19}$ rad s^{-1}

Exercise 4B

1   4 m s^{-2}
2   20.8 m s^{-2} (3 s.f.)
3   **a**   5 rad s^{-1} **b**   15 m s^{-1}
4   **a**   12.9 rad s^{-1} (3 s.f.)   **b**   7.75 m s^{-1} (3 s.f.)
5   2.14 m s^{-2} (3 s.f.)
6   0.283 rad s^{-1} (3 s.f.)
7   0.72 N
8   48.6 N
9   **a**   0.588 N (3 s.f.)   **b**   4.5 N
10   **a**   0.24 m s^{-1} **b**   0.0072 N
11   0.0294 (3 s.f.)
12   3.13 rad s^{-1} (3 s.f.)
13   0.157 (3 s.f.)
14   0.233 rad s^{-1} (3 s.f.)
15   **a**   320 N (3 s.f.)   **b**   0.000153 (3 s.f.)
16   **a**   2.42 rad s^{-1} (3 s.f.)
     b   No, because it is the minimum possible value for W. If the speed or the coefficient of friction reduced at all, the people would slip down the cylinder.
17   1.4 m s^{-1}
18   **a** $\mu > \dfrac{v^2}{gR}$
     b   Model assumes that the tyres all experience the same friction.
19   0.322 m (3 s.f.)
20   $\omega \leqslant \dfrac{\sqrt{197}}{10}$

Challenge

a $y = \dfrac{q}{p^2} x^2$

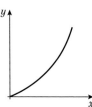

b   Acceleration is $2q$ in the positive y-direction. Speed at the origin is p.
c $y = R - \sqrt{R^2 - x^2}$
d $R = \dfrac{p^2}{2q}$
e $2q$
f   The acceleration of P and Q are equal.

Exercise 4C

1   18.4 N (3 s.f.), 4.52 rad s^{-1} (3 s.f.)
2   10.3 N (3 s.f.), 4.43 rad s^{-1} (3 s.f.)
3   23.7 N (3 s.f.), 60° (nearest degree)
4   73.5 N, 0.6 m
5   $T = ml\omega^2$
6   Let the tension in the string be T. The angle between the string and the vertical is θ, and the radius of the circle is r.
     R($\rightarrow$): $T\sin\theta = mr\omega^2$
     R($\rightarrow$): $T\cos\theta = mg$
     Dividing the horizontal component by the vertical component
     $\tan\theta = \dfrac{mr\omega^2}{mg}$
     $\dfrac{r}{x} = \dfrac{r\omega^2}{g}$ so, $\omega^2 x = g$

7   $\omega = \sqrt{\dfrac{g}{3}}$
8   $d = 5$ cm
9   $\omega = 1.8$ rad s^{-1}, $v = 5.4$ m s^{-1}
10   18.1 rad s^{-1} (3 s.f.)
11   9.5°
12   22.8 rad s^{-1} (3 s.f.)
13   9.44 rad s^{-1} (3 s.f.)
14   **a**   R($\rightarrow$): $R\cos\alpha = mg$
        R($\uparrow$): $R\sin\alpha = \dfrac{mv^2}{r}$
        Dividing the horizontal component by the vertical component to eliminate R:
        $\tan\alpha = \dfrac{mv^2}{rmg} = \dfrac{v^2}{rg}$
        So, $v^2 = rg\tan\alpha = \sqrt{rg\tan\alpha}$
     b   This model assumes there is no friction between the vehicle and the road.
15   2.22 m s^{-1} (3 s.f.)
16   0.4
17   42° (nearest degree)
18   29.5 m s^{-1} (3 s.f.), 9.94 m s^{-1} (3 s.f.)
19   **a**   20° (nearest degree), 20 800 N (3 s.f.); 68° (nearest degree), 52 700 N (3 s.f.)
     b   To turn in a shorter time the aircraft will need to decrease the radius of the circular arc in which it turns. Thus the angle to the horizontal and lift force must both increase.
20   R($\uparrow$): $T\sin\theta + R\sin\theta = mg \Rightarrow T + R = \dfrac{mg}{\sin\theta}$
     R($\rightarrow$): $T\cos\theta - R\cos\theta = ml\cos\theta\omega^2 \Rightarrow T - R = ml\omega^2$
     Eliminating R:
     $2T = \dfrac{mg}{\sin\theta} + ml\omega^2$
     $T = \dfrac{1}{2}m(l\omega^2 - g\operatorname{cosec}\theta)$
21   2.20 rad s^{-1} (3 s.f.)

Exercise 4D

1   **a**   3.13 m s^{-1} (3 s.f.)   **b**   17.6 N (3 s.f.)
2   **a**   3.43 m s^{-1} (3 s.f.)   **b**   19.6 N
3   **a**   2.97 m s^{-1} (3 s.f.)   **b**   15.7 N (3 s.f.)
4   **a**   2.21 m s^{-1} (3 s.f.)   **b**   5.88 N (3 s.f.)
5   **a**   8.52 m s^{-1} (3 s.f.)   **b**   46.9 N (3 s.f.)
6   **a** $v = \sqrt{u^2 - 1.4g(1 - \cos\theta)}$ **b** $u \geqslant \sqrt{2.8g}$
7   **a** $T = 4.5g\cos\theta + \dfrac{3u^2}{4} - 3g$ **b** $u > \sqrt{10g}$
8   **a**   0.27   **b**   0.26
9   **a**   0.30   **b**   0.28
     c   The particle will continue in a parabolic arc (projectile motion) in the negative x-direction, initially increasing in y before decreasing in y.
10   **a**   6.26 m s^{-1} (3 s.f.), 25.3 N (3 s.f.)
     b   5.6 m s^{-1}, 18.8 N (3 s.f.)
11   **a**   9.66 m s^{-1} (3 s.f.)   **b** g m s^{-2}
     c   45.8 N (3 s.f.)
12   39.6° (3 s.f.) to the upward vertical, $v = 2.74$ m s^{-1} (3 s.f.)
13   $\sqrt{8gr}$
14   **a** $\sqrt{\dfrac{16gr}{5}}$ **b** $\dfrac{11mg}{5}$
15   **a** $T = mg(1 + 3\sin\theta)$ **b**   19.5° (3 s.f.)
16   **a**   At point S: G.P.E. = $0.4 \times g \times 3.8 = 1.52g$, K.E. = 0
        At point P: G.P.E. = $0.4 \times g \times 4\sin\theta = 1.6g\sin\theta$,

K.E. $= \frac{1}{2} \times 0.4 \times v^2 = 0.2v^2$

By conservation of energy: $1.52g = 1.6g\sin\theta + 0.2v^2$

$\Rightarrow 0.2v^2 = 1.52g - 1.6g\sin\theta$

$\Rightarrow v^2 = 7.6g - 8g\sin\theta$

$\Rightarrow v = \sqrt{7.6g - 8g\sin\theta}$

b $3.8\,\text{m}$

c In reality there will be frictional forces acting on the handle so the height will be less than $3.8\,\text{m}$.

Exercise 4E

1 a $mg + 3mg\cos\theta$ **b** $\frac{4l}{3}$ **c** $\frac{40l}{27}$

2 a $9g\cos\theta - 6g$ **b** $48°$ (nearest degree)

   c $6.7\,\text{m}$

3 a $\frac{9rg}{4} - 2rg\cos\theta$ **b** $\frac{3}{4}$ **c** $\frac{\sqrt{3rg}}{4}$

   d $\frac{3\sqrt{rg}}{2}$ **e** $64°$ (nearest degree)

4 a $48°$ (nearest degree)

   b $\sqrt{8g}$, $74°$ (nearest degree)

5 a $\frac{a}{4}$ **b** $\sqrt{\frac{ga}{4}}$

   c $\sqrt{\frac{9ga}{4}}$, $64°$ (nearest degree)

6 a $49°$ (nearest degree)

   b The particle will fall through a parabolic arc (projectile motion) towards the surface in the positive x-direction.

7 a $10.3\,\text{m s}^{-1}$

   b At R: $\frac{1}{2} \times 2 \times v^2 = 2g(12 - 5\cos 70° - 7\cos 40°)$

     So, $v^2 = 96.58$

     Resolving towards B: $mg\cos\theta - R = \frac{mv^2}{7}$

     $R = 2g\cos 40° - \frac{2v^2}{7} = -12.6$

     $R < 0$

     This is impossible, so the particle must have lost contact with the chute.

   c In reality, energy is lost due to friction between the laundry bags and the chute.

8 a $\frac{\sqrt{17ga}}{3}$

   b Energy would be lost due to the frictional force acting on the marble, requiring a larger initial speed for the marble to leave the bowl.

Chapter review 4

1 $R(\rightarrow)$: $R\cos\theta = mg$

   $R(\uparrow)$: $R\sin\theta = \frac{mv^2}{r} = \frac{2mu^2}{3a}$

   Dividing the horizontal component by the vertical component gives $\tan\theta = \frac{2u^2}{3ag}$

   From the problem's geometry, $\tan\theta = \frac{\left(\frac{3a}{2}\right)}{\left(\frac{\sqrt{7a}}{2}\right)} = \frac{3}{\sqrt{7}}$

   So, $\frac{3}{\sqrt{7}} = \frac{2u^2}{3ag}$

   $9ag = 2\sqrt{7}\,u^2$

2 a $\frac{3mg}{2}$ **b** $\frac{mg}{2}$

3 a $\frac{13mg}{5}$ **b** $\sqrt{60gl}$

4 $108\,\text{m}$ (3 s.f.)

5 a T is the tension in string AP. S is the tension in string BP.

    The triangle is equilateral.

    $R(\rightarrow)$: $T\cos 60° = mg + S\cos 60°$

         $T - S = 2mg$

    $R(\uparrow)$: $T\cos 30° + S\cos 30° = mr\omega^2$

        $(T + S)\cos 30° = m(l\cos 30°)\omega^2$

            $T + S = ml\omega^2$

    Eliminating S to find T:

    $2T = 2mg + ml\omega^2$

      $T = \frac{m}{2}(2g + l\omega^2)$

   b $\frac{m}{2}(l\omega^2 - 2g)$

   c Both strings taut $\Rightarrow l\omega^2 - 2g > 0$, $\omega^2 > \frac{2g}{l}$

6 a Let T be the tension in the string.

    $R(\uparrow)$: $T\cos 45° = mg$

         $T = \sqrt{2}\,mg$

   b $\omega = \sqrt{\frac{\sqrt{2}g}{l}}$

7 a $6.48\,\text{N}$ **b** $25°$ (nearest degree)

8 a $4mg$ **b** $3\pi\sqrt{\frac{r}{g}}$

9 a $\frac{mv^2}{r} = \mu R = \mu mg$

       $\frac{v^2}{rg} = \mu$

     $\frac{21^2}{100 \times 9.8} = \mu$

        $\mu = 0.45$

   b $\frac{35}{136}$

10 a $\frac{\sqrt{3}m}{4}(r\omega^2 + 2g)$

   b Maximum speed gives the shortest time. At the maximum speed with the rod still on the surface of the sphere, $R = 0$. Radius of the circle is $\frac{\sqrt{3}r}{2}$.

    When $R = 0$, $T\cos\alpha = mg \Rightarrow T = \frac{mg}{\cos\alpha} = \frac{2mg}{\sqrt{3}}$

         $T\sin\alpha = m \times \frac{\sqrt{3}r}{2} = \omega^2$

    So $\frac{2mg}{\sqrt{3}} \times \frac{1}{2} = m \times \frac{\sqrt{3}r}{2} = \omega^2$

         $\omega^2 = \sqrt{\frac{2g}{3r}}$

    Time for one revolution $= \frac{2\pi}{\omega} = \pi\sqrt{\frac{4 \times 3r}{2g}} = \pi\sqrt{\frac{6r}{g}}$

   c i The minimum period decreases.

      ii The minimum period increases.

11 a Let F be the force due to friction.

    Let R be the normal reaction force.

    $R(\rightarrow)$: $R = mg$

    $R(\uparrow)$: $F = mr\omega^2$

    If P does not slip, then $\mu R \geqslant mr\omega^2$

    $\frac{3}{7}mg \geqslant \frac{3a}{5}m\omega^2 \Rightarrow \omega^2 \leqslant \frac{5g}{7a}$

   b $\frac{5g}{42a} \leqslant \omega^2 \leqslant \frac{65g}{42a}$

12 a $\frac{4}{3}ga + 2ga\sin\theta$ **b** $mg\left(\frac{4}{3} + 3\sin\theta\right)$

   c $206°$

   d $v = 0$ before the particle reaches the top of the circle.

13 $\sqrt{2.6g}$

14 a $\dfrac{ga}{2}(5 - 4\cos\theta)$

    b Resolving towards the centre O:

    $mg\cos\theta - R = \dfrac{mv^2}{r}$

    $mg\cos\theta - R = \dfrac{mg}{2}(5 - 4\cos\theta)$

    Substituting $\cos\theta = 0.9$

    $R = 0.9mg - \dfrac{mg}{2}(5 - 3.6)$

    $R = 0.2mg$

    $R > 0 \Rightarrow P$ is still on the hemisphere

    c i $\dfrac{5}{6}$ **ii** $\sqrt{\dfrac{5ga}{6}}$

    d $\sqrt{\dfrac{5ga}{2}}$ **e** 61° (nearest degree)

15 a $u^2 - \dfrac{16}{5}gr$ **b** $\dfrac{mg}{5}$

    c $\sqrt{\dfrac{19gr}{5}}$ **d** $\sqrt{\dfrac{73gr}{15}}$

16 a $\dfrac{u^2 + 2ag}{3ag}$ **b** 34° (nearest degree)

Challenge

a At point (x, x^2), $\dfrac{dy}{dx} = 2x$

  R($\uparrow$): $R\cos\theta = mg$ (1)

  R($\rightarrow$): $R\sin\theta = mx\omega^2$ (2)

  (2) $\div$ (1): $\tan\theta = \dfrac{x\omega^2}{g}$ (3)

  $\tan\theta = \dfrac{dy}{dx} = 2x$

  $\Rightarrow 2x = \dfrac{x\omega^2}{g} \Rightarrow 2g = \omega^2 \Rightarrow \omega = \sqrt{2g}$

  Hence ω is independent of the vertical height.

b From (3), $\omega^2 = \dfrac{g\tan\theta}{x}$

  For ω to be independent of $x \Rightarrow \dfrac{g\tan\theta}{x} = k$ for constant k

  $\Rightarrow \tan\theta = ax$ for constant a

  $\dfrac{dy}{dx} = \tan\theta = ax \Rightarrow y = \dfrac{1}{2}ax^2 + b$

  Hence $f(x) = px^2 + q$ for for constants p and q

CHAPTER 5

Prior knowledge check

1 $\dfrac{26}{3}$

2 $A = 8$ and $B = 4$

3 $\dfrac{262}{91}$

Exercise 5A

1 $\left(\dfrac{2}{3}, 2\right)$

2 $(1.5, 3.6)$

3 $(2.4, 0.75)$

4 $\left(\dfrac{14}{25}, \dfrac{23}{35}\right)$

5 $\left(\dfrac{3}{5}a, 0\right)$

6 $\left(\dfrac{\pi}{2}, \dfrac{\pi}{8}\right)$

7 $\left(\dfrac{1 - \ln 2}{\ln 2}, \dfrac{1}{4\ln 2}\right)$

8 $\left(\dfrac{4\sqrt{2}\,r}{3\pi}, 0\right)$

9 $\left(\dfrac{16}{15}, \dfrac{64}{21}\right)$

10 $\left(\dfrac{254}{55}, \dfrac{15}{11}\right)$

11 $(1.01, 0)$ (3 s.f.)

12 $(1.34, 0.206)$ (3 s.f.)

Challenge
$(0, 0), (4, 0), (4, 2), (4, -2), \left(\dfrac{19}{10}, \sqrt{\dfrac{19}{10}}\right)$ and $\left(\dfrac{19}{10}, -\sqrt{\dfrac{19}{10}}\right)$

Exercise 5B

1 $(2, 0)$ **2** $(1, 0)$

3 $(\pi, 0)$ **4** $(0, -3)$

5 $\left(\dfrac{5}{6}, 0\right)$

6 $\left(2\dfrac{2}{3}, 0\right)$

7 $\left(\dfrac{35}{48}, 0\right)$

8 $(1.65, 0)$ (3 s.f.)

9 $\left(\dfrac{1}{2}\dfrac{(e^2 + 1)}{(e^2 - 1)}, 0\right)$

10 $(0, 1.34)$ (3 s.f.)

11 $(1.01, 0)$ (3 s.f.)

12 1.46 (3 s.f)

13 3.04 cm

14 $\dfrac{r}{3}$ above the base

15 $\dfrac{5}{6}$ cm above the base

16 The arc of the circle $x^2 + y^2 = a^2$, $a - h \leqslant x \leqslant a$ is rotated about the x-axis.

$\bar{x} = \dfrac{\pi \int\limits_{a-h}^{a}(a^2 - x^2)x\,dx}{\pi \int\limits_{a-h}^{h}(a^2 - x^2)\,dx} = \dfrac{\left[\dfrac{1}{2}a^2x^2 - \dfrac{1}{4}x^4\right]_{a-h}^{a}}{\left[a^2x - \dfrac{1}{2}x^3\right]_{a-h}^{a}}$

$= \dfrac{\dfrac{1}{4}a^4 - \dfrac{1}{2}a^2(a - h)^2 + \dfrac{1}{4}(a - h)^4}{\dfrac{2}{3}a^3 - a^2(a - h) + \dfrac{1}{3}(a - h)^2}$

$= \dfrac{\dfrac{1}{4}(a^2 - (a - h)^2)^2}{\dfrac{1}{3}(2a^2 - 3a^2(a - h) + (a - h)^3)}$

$= \dfrac{3}{4}\dfrac{(2ah - h^2)^2}{3h^2a - h^3}$

$= \dfrac{3}{4}\dfrac{(2a - h)^2}{(3a - h)}$

$\therefore$ Distance of centre of mass from base of cap (i.e. $x = a - h$) is

$\dfrac{3(2a - h)^2}{4(3a - h)} - (a - h) = \dfrac{3(2a - h)^2 - 4(3a - h)(a - h)}{4(3a - h)}$

$= \dfrac{4ah - h^2}{4(3a - h)}$

$= \dfrac{h(4a - h)}{4(3a - h)}$

17 a $\bar{y} = 3$ **b** 4.18 cm

18 $k = \dfrac{7}{5}$

Online Worked solutions are available in SolutionBank.

19 a $\dfrac{23h}{44}$

b On the common axis of symmetry of the cone and cylinder, a distance $\dfrac{23h}{44}$ below the top plane face of the cylinder and a distance $\dfrac{r}{11}$ to the left of the vertical axis of the cylinder.

20 a 4.07 cm (3 s.f.) **b** 4.17 cm (3 s.f.)

Challenge

$\dfrac{R}{2}$ from the plane face

Exercise 5C

1 The centre of mass lies on the axis of symmetry at a point 0.42 cm below the base of the cone.

2 Centre of mass is on the axis of symmetry at a distance 5.33 cm away from base of the cylinder.

3 a A square based pyramid has base area A and height h, the centre of mass is on the axis of symmetry.

Volume $= \dfrac{1}{3}Ah$

Mass $= \dfrac{1}{3}Ah\rho$

Take a slice of thickness δx at a distance x_i from vertex. The base of the slice is an enlargement of the base of the pyramid with scale factor $\dfrac{x_i}{h}$

Ratio of areas is $\left(\dfrac{x_i}{h}\right)^2$

Area of base of slice is $\dfrac{x_i^2}{h^2}A$

Mass of slice $m_i = \delta x$

$\lim\limits_{\delta x \to 0}\sum\limits_{x=0}^{h} m_i x_i = \int\limits_0^h \dfrac{x^3}{h^2}A\rho\,dx = \dfrac{1}{4}h^2 A\rho$

$\bar{x} = \dfrac{\sum m_i x_i}{M} = \dfrac{\frac{1}{4}h^2 A\rho}{\frac{1}{3}Ah\rho} = \dfrac{3}{4}h$

The centre of mass lies on the line of symmetry at a distance $\dfrac{3}{4}h$ from the vertex, or $\dfrac{h}{4}$ from base.

b 3.39 cm (3 s.f.) below O

4 a $\dfrac{h}{4}$ where h is the height of the tetrahedron.

b $\dfrac{3}{8}\sqrt{6}$ m below O

5 a 1.96 cm (3 s.f.) **b** 0.48 cm below O (3 s.f.)

6 a

Shape	Mass	Vertical distance from O to CoM
Cylinder	$32\pi r^3$	r
Hemisphere	$\dfrac{2\pi r^3}{3}$	$\dfrac{3r}{8}$
Composite body	$32\pi r^3 - \dfrac{2\pi r^3}{3} = \dfrac{94\pi r^3}{3}$	$\bar{x}$

$32\pi r^3(r) - \dfrac{2\pi r^3}{3}\left(\dfrac{3r}{8}\right) = \dfrac{94\pi r^3}{3}(\bar{x})$

$\dfrac{127\pi r^4}{4} = \dfrac{94\pi r^3}{3}\bar{x} \Rightarrow \bar{x} = \dfrac{381r}{376}$

b i $\dfrac{387r}{392}$ **ii** $\dfrac{r}{49}$

Challenge

$a = -2.5$ and $b = 3$

Chapter review 5

1 a $V = \int \pi y^2 \, dx = \pi \int\limits_0^4 4x \, dx$

$= \pi[2x^2]_0^4$

$= 32\pi$

b $\bar{x} = \dfrac{8}{3}$

2 a $V = \int \pi y^2 \, dx = \pi \int\limits_1^2 \dfrac{1}{x^2} \, dx$

$= \pi\left[\dfrac{-1}{x}\right]_1^2$

$= \dfrac{\pi}{2}m^2$

b 39 cm to the nearest cm

3 34 cm

4 $\dfrac{9r}{10}$

5 0.391

6 $\left(\dfrac{2\ln 2 - \frac{3}{4}}{2\ln 2 - 1}, 0\right)$ or (1.65, 0)

7 $x = \dfrac{\pi^2 - 4}{4\pi}$

Chapter 6

Prior knowledge check

1 $T_1 = \dfrac{\sqrt{2}}{\sqrt{3}+1}mg$, $T_2 = \dfrac{\sqrt{3}-1}{\sqrt{3}+1}mg$

2 a North: −2.64 N, East: 4.56 N

b Resultant is 5.27 N

Exercise 6A

1 $\theta = 68°$ (to the nearest degree)

2 $\theta = 40°$ (to the nearest degree)

3 $\theta = 63°$ (to the nearest degree)

4 a 2.78 cm from large circular face

b $\alpha = 51°$ (to the nearest degree)

5 $T_1 = 13.3$ N, $T_2 = 6.34$ N

6 a $T_1 = 1.07 \times 10^{-3}$ N, $T_2 = 1.21 \times 10^{-3}$ N

b 46.7° (3 s.f.)

7 a In equilibrium the centre of mass G lies below the point of suspension S. Let distance $SG = x$.

O is the centre of the base of the cone and V is its vertex.

A is the point on the base connected to the string and B is the point on the line SG a distance r from G.

$\tan\theta = \dfrac{x}{3r}$ (from triangle VSG)

Also $\tan\theta = \dfrac{x-r}{r}$ (from triangle ABS)

$\dfrac{x}{3r} = \dfrac{x-r}{r}$

$x = 3x - 3r$

$2x = 3r$

$x = \dfrac{3r}{2}$

$\tan\theta = \dfrac{1}{2}$

b $\dfrac{\sqrt{5}\,mg}{2}$

8 22.8° (3 s.f.)

Exercise 6B

1 4.77 cm (3 s.f.)

2 a 31° (to the nearest degree) **b** $\mu = \dfrac{3}{5}$

3 a 30° **b** 35 cm (2 s.f.)

4 a $\dfrac{2Mg}{2 + \sqrt{3}}$ **b** $\dfrac{1}{2}$

5 a Let the height of the small cone be h.
Using similar triangles:

$$\frac{h}{h + 2r} = \frac{r}{2r}$$
$$2h = h + 2r$$
$$h = 2r$$

Shape	Mass	Mass ratio	Distance of CoM from centre axis
Large cone	$\rho\dfrac{1}{3}\pi(2r)^2\,4r$	8	r
Small cone	$\rho\dfrac{1}{3}\pi r^2 \times 2r$	1	$2r + \dfrac{2r}{4} = \dfrac{5r}{2}$
Frustum	$\rho\dfrac{1}{3}\pi \times 14r^3$	7	$\bar{x}$

Take moments about centre axis:

$$8r - \frac{5r}{2} = 7\bar{x}$$
$$\bar{x} = \frac{11r}{14}$$

b i Yes **ii** No

c As the angle of the slope is 40°, limiting friction would imply $\mu = \tan 40°$.
No slipping implies that $\mu \geqslant 0.839$ (3 s.f.)

6 a Consider the cube in equilibrium, on the point of toppling, so the reaction force acts through the bottom corner A.
R($\rightarrow$): $P - F = 0 \Rightarrow F = P$
R($\uparrow$): $R - W = 0 \Rightarrow R = W$
Moments about A: $P \times 4a = W \times 3a$
$\Rightarrow P = \dfrac{3}{4}W$

If equilibrium is broken by toppling:
$P = \dfrac{3}{4}W \Rightarrow F = \dfrac{3}{4}W$
But $F < \mu R$
$\Rightarrow \dfrac{3}{4}W < \mu W \Rightarrow \mu > \dfrac{3}{4}$ is the condition for toppling.
If $\mu < \dfrac{3}{4}$ then the cube will be on the point of slipping when $F = \mu R$.

b $2a$

7 a 25.6° **b** $0 < k < 6$

8 a

Shape	Mass	Mass ratio	Distance of CoM from O
Large cone	$\dfrac{1}{3}\pi\rho(2r^2)2h$	8	$\dfrac{2h}{4}$
Small cone	$\dfrac{1}{3}\pi\rho r^2 h$	1	$h + \dfrac{h}{4}$
Frustum	$\dfrac{1}{3}\pi\rho(8r^2h - r^2h)$	7	$\bar{x}$

The centre of the base is the point O.
The radius of the smaller cone is obtained by similar triangles.

Moments: $8 \times \dfrac{2h}{4} - 1 \times \dfrac{5h}{4} = 7\bar{x}$

$$\frac{11h}{4} = 7\bar{x} \Rightarrow \bar{x} = \frac{11}{28}h$$

b $\theta = 38°$ (to the nearest degree)

9 a $P > \mu Mg$ **b** $P > \dfrac{3}{8}Mg$

c i Slide **ii** Topple **iii** Slide and topple

10 a $P = \dfrac{9g}{40}$

b $P = \dfrac{3g}{10}$

c $\dfrac{9g}{40} < \dfrac{3g}{10}$

Chapter review 6

1 a 0.71 (2 s.f.)

b The centre of mass of the body is at C which is always directly above the contact point.

2 a $\bar{y} = \dfrac{\rho\int\dfrac{1}{2}y^2\,dx}{\rho\int y\,dx} = \dfrac{\dfrac{1}{2}\displaystyle\int_0^4\dfrac{x^2}{16}(16 - 8x + x^2)dx}{\dfrac{1}{4}\displaystyle\int_0^4 4x - x^2\,dx}$

$= 2\dfrac{\left[\dfrac{1}{3}x^3 - \dfrac{1}{8}x^4 + \dfrac{1}{80}x^5\right]_0^4}{\left[2x^2 - \dfrac{1}{3}x^3\right]_0^4} = \dfrac{2}{5}$

b $\theta = 79°$ (to the nearest degree)

3 a Take the diameter as the y-axis and the midpoint of the diameter as the origin.
Then $M\bar{x} = \rho\int 2yx\,dx$ where
$M = \dfrac{1}{2}\rho\pi(2a)^2$ and where $x^2 + y^2 = (2a)^2$

$2\rho\pi a^2\bar{x} = \rho\displaystyle\int_0^{2a} 2x\sqrt{4a^2 - x^2}\,dx$

$= \dfrac{-2\rho}{3}[(4a^2 - x^2)^{\frac{3}{2}}]_0^{2a}$

$2\rho\pi a^2\bar{x} = \dfrac{2\rho}{3} \times 8a^3$

$\bar{x} = \dfrac{16}{3}a^3 \div 2\pi a^2$

$= \dfrac{8a}{3\pi}$

b

Shape	Mass	Mass ratio	Distance of CoM from AB
Large semicircle	$2\pi\rho a^2$	4	$\dfrac{8a}{3\pi}$
Semicircle diameter AD	$\dfrac{1}{2}\pi\rho a^2$	1	$\dfrac{4a}{3\pi}$
Semicircle diameter OB	$\dfrac{1}{2}\pi\rho a^2$	1	$\dfrac{4a}{3\pi}$
Remainder	$\pi\rho a^2$	2	$\bar{x}$

Moments: $4 \times \dfrac{8a}{3\pi} - 1 \times \dfrac{4a}{3\pi} - 1 \times \dfrac{4a}{3\pi} = 2\bar{x}$

$\dfrac{24a}{3\pi} = 2\bar{x}$

$\bar{x} = \dfrac{4a}{\pi}$

c The distance from OC is a
The distance from OB is $\dfrac{2a}{\pi}$

d 78° (to the nearest degree)

Online — Worked solutions are available in SolutionBank.

4 a

Shape	Mass	Mass ratio	Distance of CoM from O
Cylinder	$\pi\rho r^2 h$	h	$\dfrac{-h}{2}$
Hemisphere	$\dfrac{2}{3}\pi\rho(3r)^3$	$18r$	$\dfrac{3}{8}(3r)$
Mushroom	$\pi\rho r^2(h+18r)$	$h+18r$	0

Moments: $-h \times \dfrac{h}{2} + 18r \times \dfrac{3}{8} \times 3r = 0$

$\dfrac{h^2}{2} = \dfrac{81r^2}{4}$

$h = r\sqrt{\dfrac{81}{2}}$

b $\theta = 9°$ (to the nearest degree)

5 a $V = \pi \int_0^a 4ax\,dx$

$\quad = \pi[2ax^2]_0^a$

$\quad = 2\pi a^3$

b $\bar{x} = \dfrac{\pi \int_0^a 4ax^2\,dx}{2\pi a^3}$

$\quad = \pi\dfrac{\left[\dfrac{4ax^3}{3}\right]_0^a}{2\pi a^3}$

$\quad = \dfrac{2}{3}a$

c $\rho_1 : \rho_2 = 6 : 1$

d As centre of mass is at centre of hemisphere this will always be above the point of contact with the plane. (Tangent–radius property).

6 a

Shape	Mass	Mass ratio	Distance of CoM from AB
Cylinder	$\pi\rho(2r)^2 \times 3r$	$12r$	$\dfrac{3r}{2}$
Cone	$\dfrac{1}{3}\pi\rho r^2 \times h$	$\dfrac{1}{3}h$	$\dfrac{1}{4}h$
Remainder	$\pi\rho\left(12r^3 - \dfrac{1}{3}r^2 h\right)$	$12r - \dfrac{1}{3}h$	$\vec{x}$

Moments: $\left(12r - \dfrac{1}{3}h\right)\bar{x} = 12r \times \dfrac{3r}{2} - \dfrac{h}{3} \times \dfrac{h}{4}$

$\left(12r - \dfrac{1}{3}h\right)\bar{x} = 18r^2 - \dfrac{h^2}{12}$

$\bar{x} = \dfrac{18r^2 - \dfrac{h^2}{12}}{12r - \dfrac{h}{3}}$

$\bar{x} = \dfrac{216r^2 - h^2}{4(36r - h)}$

b $\theta = 38°$ (to the nearest degree)

7 a $0.265\,m$ (3 s.f.)   **b** $0.478mg$ (3 s.f.)

8 a 1.11 (3 s.f.)   **b** $14.4°$ (3 s.f.)

Challenge

a

Shape	Mass	Mass ratio	Distance of CoM from O
Cone	$\dfrac{1}{3}\pi\rho r^2 h$	h	$\dfrac{h}{4}$
Hemisphere	$\dfrac{2}{3}\pi\rho r^3$	$2r$	$\dfrac{-3r}{8}$
Toy	$\dfrac{1}{3}\pi\rho(r^2 h + 2r^3)$	$h+2r$	$\vec{x}$

Moments: $(h+2r)\bar{x} = h \times \dfrac{h}{4} + 2r\left(\dfrac{-3r}{8}\right)$

$(h+2r)\bar{x} = \dfrac{h^2}{4} - \dfrac{3r^2}{4}$

$\bar{x} = \dfrac{h^2 - 3r^2}{4(h+2r)}$

b i Fall over
ii Return to vertical position
iii Remain in new position

Review exercise 2

1 $3.67\,\text{m s}^{-1}$ (3 s.f.)

2 a $T = 5.5\,\text{N}$ (2 s.f.)   **b** $\theta = 26°$ (nearest degree)

3 $190\,\text{m}$ (2 s.f.)

4 $24\,\text{m s}^{-1}$ (2 s.f.)

5 a $\alpha = 70.5°$ (3 s.f.)   **b** $l = \dfrac{3}{2k}$

6 a $2mg$ **b** $2\pi\sqrt{\dfrac{r}{2g}}$

7 a $\tan 60° = \dfrac{r}{\frac{h}{2}} \Rightarrow r = \dfrac{h}{2}\tan 60° = \dfrac{\sqrt{3}}{2}h$

b Tension in $AP = mg + \dfrac{1}{2}mh\omega^2$ and tension in $BP = \dfrac{1}{2}mh\omega^2 - mg$

c Tension in $BP > 0 \Rightarrow \omega > \sqrt{\dfrac{2g}{h}}$

As $T = \dfrac{2\pi}{\omega}$, $T < 2\pi\sqrt{\dfrac{h}{2g}} \Rightarrow T < \pi\sqrt{\dfrac{2h}{g}}$

8 $2a$

9 a $R(\uparrow): R - mg = 0 \Rightarrow R = mg$

$R(\leftarrow): F = mr\omega^2 = m\left(\dfrac{4}{3}a\right)\omega^2$

As P remains at rest $F \leqslant \mu R$

$\Rightarrow m\left(\dfrac{4}{3}a\right)\omega^2 \leqslant \dfrac{3}{5}mg \Rightarrow \omega^2 \leqslant \dfrac{9g}{20a}$

b $\omega^2_{\max} = \dfrac{19g}{20a}$ and $\omega^2_{\min} = \dfrac{g}{20a}$

10 a $u = \sqrt{\dfrac{gl}{2}}$ **b** $\dfrac{5l}{6}$

11 a Use conservation of energy:

$\dfrac{1}{2}m(u^2 - v^2) = mgl(1 - \cos\theta)$

$\Rightarrow v^2 = u^2 - 2gl(1 - \cos\theta) = 3gl - 2gl + 2gl\cos\theta$

$v^2 = gl + 2gl\cos\theta$

Resolve along the string:

$T - mg\cos\theta = \dfrac{mv^2}{l} = \dfrac{mgl + 2mgl\cos\theta}{l}$

$\Rightarrow T = mg(1 + 3\cos\theta)$

b $v = \sqrt{\dfrac{gl}{3}}$ **c** $\dfrac{40l}{27}$

12 a $\sqrt{\dfrac{2gl}{3}}$

b Resolve along the string: $T - mg\cos\theta = \dfrac{mv^2}{l}$ (1)

Conservation of energy:

$\dfrac{1}{2}mu^2 - \dfrac{1}{2}mv^2 = mgl(1 - \cos\theta)$ (2)

Solving (1) and (2) simultaneously: $T = \dfrac{mg}{3}(9\cos\theta - 4)$

c $\dfrac{2mg}{3} \leqslant T \leqslant \dfrac{5mg}{3}$

13 a $\dfrac{1}{2}mv^2 = mg(a\cos\alpha - a\cos\theta) \Rightarrow v^2 = 2ga(\cos\alpha - \cos\theta)$

b $\theta = 60°$ or $\dfrac{\pi}{3}$ radians   **c** $\sqrt{\dfrac{7ga}{2}}$

14 a $v = 8$ **b** $m = 6$ **c** $1300\,\text{N}$ (2 s.f.)

15 a $30\,\text{m s}^{-1}$ (2 s.f.)   **b** $1900\,\text{N}$

c $28\,\text{m s}^{-1}$ (2 s.f.)   **d** $560\,\text{N}$ (2 s.f.)

e Lower speed at $C \Rightarrow$ the normal reaction is reduced.

16 a $\dfrac{2}{3} + \dfrac{u^2}{3ag}$ **b** $34°$ (nearest degree)

17 $\dfrac{8}{3}$ from O

18 a The centre of mass lies on the x-axis from symmetry.

An elemental strip of area is $2y\delta x$

The boundary of the semicircle has equation $x^2 + y^2 = a^2, 0 \leqslant x \leqslant a$

$\bar{x} = \dfrac{\rho \int 2x(a^2 - x^2)^{\frac{1}{2}}dx}{\rho \times \dfrac{\pi a^2}{2}} = \dfrac{2}{\pi a^2}\left[-\dfrac{2}{3}(a^2 - x^2)^{\frac{3}{2}}\right]_0^a = \dfrac{4a}{3\pi}$

b

Shape	Mass	Mass ratio	Distance of CoM from O
Semicircle radius a	$\dfrac{1}{2}\pi\rho a^2$	a^2	$\dfrac{4a}{3\pi}$
Semicircle radius b	$\dfrac{1}{2}\pi\rho b^2$	b^2	$\dfrac{4b}{3\pi}$
Remainder	$\dfrac{1}{2}\pi\rho(a^2 - b^2)$	$a^2 - b^2$	$\bar{x}$

Moments about O : $a^2 \times \dfrac{4a}{3\pi} - b^2 \times \dfrac{4b}{3\pi} = (a^2 - b^2)\bar{x}$

so $\bar{x} = \dfrac{4}{3\pi}\dfrac{(a^3 - b^3)}{(a^2 - b^2)} = \dfrac{4}{3\pi}\dfrac{(a - b)(a^2 + ab + b^2)}{(a - b)(a + b)}$

$= \dfrac{4(a^2 + ab + b^2)}{3\pi(a + b)}$

c $\dfrac{2a}{\pi}$

19 a $\bar{x} = \dfrac{\int \pi xy^2\,dx}{\int \pi y^2\,dx} = \dfrac{\dfrac{1}{4}\int_1^2 x^{-3}\,dx}{\dfrac{1}{4}\int_1^2 x^{-4}\,dx} = \dfrac{\left[-\dfrac{1}{2}x^{-2}\right]_1^2}{\left[-\dfrac{1}{3}x^{-3}\right]_1^2} = \dfrac{9}{7}$

The centre of mass is $\left(\dfrac{9}{7} - 1\right) = \dfrac{2}{7}$ m from the larger plan face.

b $\dfrac{29}{30}$ m or 0.967 m

20 a $\bar{x} = \dfrac{\int_0^R \pi x(R^2 - x^2)\,dx}{\dfrac{2}{3}\pi R^3} = \dfrac{\left[\pi R^2\dfrac{x^2}{2} - \dfrac{\pi x^4}{4}\right]_0^R}{\dfrac{2}{3}\pi R^3} = \dfrac{\dfrac{1}{4}\pi R^4}{\dfrac{2}{3}\pi R^3}$

$= \dfrac{3}{8}R$

b

Shape	Mass	Mass ratio	Position of CoM
Cone	$\dfrac{1}{3}\pi\rho a^3 k$	k	$\dfrac{3}{4}ka$
Hemisphere	$\dfrac{2}{3}\pi\rho a^3$	2	$ka + \dfrac{3}{8}a$
Top	$\dfrac{1}{3}\pi\rho a^3(k + 2)$	$k + 2$	$\bar{x}$

Taking moments about V:

$k\left(\dfrac{3}{4}ka\right) + 2\left(ka + \dfrac{3}{8}a\right) = (k + 2)\bar{x} \Rightarrow \bar{x} = \dfrac{(3k^2 + 8k + 3)a}{4(k + 2)}$

c $k = \sqrt{3}$

21 a

Shape	Mass	Mass ratio	Distance of CoM from O
Large hemisphere	$\dfrac{2}{3}\pi\rho a^3$	8	$\dfrac{3}{8}a$
Small hemisphere	$\dfrac{2}{3}\pi\rho\left(\dfrac{a}{2}\right)^3$	1	$\dfrac{3}{16}a$
Remainder	$\dfrac{2}{3}\pi\rho\dfrac{7a^3}{8}$	7	$\bar{x}$

Taking moments about O: $8 \times \dfrac{3}{8}a - 1 \times \dfrac{3}{16}a = 7\bar{x}$

$\Rightarrow \bar{x} = \dfrac{45a}{112}$

b $k = \dfrac{2}{7}$

22 a $\dfrac{8\pi}{5}$

b $\bar{x} = \dfrac{\pi\int_0^2 \dfrac{1}{4}x(x - 2)^4\,dx}{\dfrac{8\pi}{5}} = \dfrac{5}{8\pi} \times \dfrac{\pi}{4}\int_{-2}^0 (u + 2)u^4\,du$

$= \dfrac{5}{32}\int_{-2}^0 u^5 + 2u^4\,du = \dfrac{5}{32}\left[\dfrac{1}{6}u^6 + \dfrac{2}{5}u^5\right]_{-2}^0$

$= \dfrac{5}{32}\left(0 - \left(-\dfrac{64}{30}\right)\right) = \dfrac{1}{3}$

c $\dfrac{59W}{12}$

23 a

Shape	Mass	Mass ratio	Distance of CoM from O
Cylinder	$\pi\rho r^2 h$	1	$\dfrac{h}{2}$
Cone	$\dfrac{1}{3}\pi\rho r^2\left(\dfrac{h}{2}\right)$	$\dfrac{1}{6}$	$h - \dfrac{1}{4}\left(\dfrac{h}{2}\right)$
Ornament	$\dfrac{5}{6}\pi\rho r^2 h$	$\dfrac{5}{6}$	$\bar{x}$

Taking moments about O, centre of plane base:

$1 \times \dfrac{h}{2} - \dfrac{1}{6} \times \dfrac{7h}{8} = \dfrac{5}{6}\bar{x} \Rightarrow \dfrac{5}{6}\bar{x} = \dfrac{17h}{48} \Rightarrow \bar{x} = \dfrac{17h}{40}$

b $\alpha = 66.5°$ (1 d.p.)

24 a

Shape	Mass	Distance of CoM from O
Circular disc	$\pi\rho a^2$	0
Hemispherical bowl	$2\pi\rho a^2$	$\dfrac{1}{2}a$
Closed container	$3\pi\rho a^2$	$\bar{x}$

Taking moments about O:

$0 + 2\pi\rho a^2 \times \dfrac{a}{2} = 3\pi\rho a^2\bar{x} \Rightarrow \bar{x} = \dfrac{a}{3}$

Online   Worked solutions are available in SolutionBank.

b 56° (nearest degree)

25 a

Shape	Mass	Distance of CoM from C
H	$8M$	$\dfrac{3a}{8}$
K	M	$-\dfrac{3a}{16}$
S	$9M$	$\bar{x}$

Taking moments about C:

$$8M \times \frac{3a}{8} - M \times \frac{3a}{16} = 9M\bar{x} \Rightarrow 3a - \frac{3a}{16} = 9\bar{x}$$

$$\Rightarrow \bar{x} = \frac{45a}{16} \div 9 = \frac{5a}{16}$$

b $\dfrac{16}{45}$

26 a

Shape	Mass	Mass ratio	Position of CoM
Hemisphere	$4\pi\rho r^3$	$4r$	$\dfrac{5r}{8}$
Cylinder	$\pi\rho r^2 h$	h	$\dfrac{h}{2} + r$
Toy	$\pi\rho r^2(4r + h)$	$4r + h$	$\bar{x}$

Taking moments about O:

$$4r \times \frac{5r}{8} + h\left(\frac{h}{2} + r\right) = (4r + h)\bar{x}$$

$$\Rightarrow \bar{x} = \frac{\dfrac{5r^2}{2} + \dfrac{h^2}{2} + rh}{4r + h} \text{ so } d = \frac{h^2 + 2rh + 5r^2}{2(h + 4r)}$$

b $h = \sqrt{3}\,r$

27 a

Shape	Mass	Distance of CoM from AB
Hemisphere	M	$\dfrac{3r}{8}$
Cone	m	$-\dfrac{3r}{4}$
Toy	$m + M$	$\bar{x}$

Taking moments about AB:

$$(m + M)\bar{x} = \frac{3Mr}{8} - \frac{3mr}{4}$$

$$\Rightarrow \bar{x} = \frac{3(M - 2m)r}{8(M + m)}$$

b Let α = angle between OA and axis of cone

No equilibrium $\Rightarrow \bar{x} > r\tan\alpha$

$$\tan\alpha = \frac{r}{3r} \Rightarrow \bar{x} > \frac{1}{3}r$$

So $\dfrac{3(M - 2m)r}{8(M + m)} > \dfrac{1}{3}r$

$9(M - 2m) > 8(M + m) \Rightarrow M > 26m$

28 a $\bar{x} = \dfrac{\dfrac{1}{2}\int_0^\pi \sin^2 x \, dx}{\int_0^\pi \sin x \, dx} = \dfrac{\dfrac{1}{4}\int_0^\pi 1 - \cos 2x \, dx}{\int_0^\pi \sin x \, dx}$

$= \dfrac{\dfrac{1}{4}\left[x - \dfrac{1}{2}\sin 2x\right]_0^\pi}{[-\cos x]_0^\pi} = \dfrac{\dfrac{\pi}{4}}{2} = \dfrac{\pi}{8}$

b 76° (nearest degree)

29 a $\dfrac{51a}{64}$ or $0.797a$ (3 s.f.)

b $\alpha = 70.6°$ (3 s.f.)

c $\beta = 38.7°$ (3 s.f.)

30 a

Shape	Mass	Distance of CoM from O
Hemisphere	$2M$	$h + \dfrac{3r}{8}$
Cylinder	$3M$	$\dfrac{h}{2}$
Combined solid	$5M$	$\bar{x}$

Taking moments about O:

$$2M\left(h + \frac{3r}{8}\right) + 3M \times \frac{h}{2} = 5M\bar{x}$$

$$\Rightarrow \frac{7h}{2} + \frac{3r}{4} = 5\bar{x}$$

$$\Rightarrow \bar{x} = \frac{14h + 3r}{20}$$

b $\dfrac{6}{7}r$

31 a $\dfrac{r}{2}$ below O

b 40.6°

c The toy will not topple.

Challenge

1 $e = \dfrac{1}{2}$

2 a Reaction of ring on ball $= \dfrac{mv^2}{R}$

Using $F = ma$ with frictional force:

$$\frac{m\mu v^2}{R} = -m\frac{dv}{dt}$$

$$\frac{1}{v^2}\frac{dv}{dt} = \frac{-\mu}{R}$$

$$\frac{-1}{v} = \frac{-\mu t}{R} + c$$

When $t = 0$, $v = u \Rightarrow c = \dfrac{-1}{u}$

$$\frac{-1}{v} = \frac{-\mu t}{R} - \frac{1}{u} \Rightarrow v = \frac{1}{\dfrac{\mu t}{R} + \dfrac{1}{u}} = \frac{uR}{R + u\mu t}$$

b $\dfrac{e^{\frac{\pi}{2}} - 1}{20}$ s or 0.191 s (3 s.f.)

Exam practice

1 a $v^2 = 64 - 4x^2$

b 4 m

2 a $V = \pi\displaystyle\int_0^{\frac{\pi}{4}} \cos^2 x \, dx = \frac{\pi}{2}\int_0^{\frac{\pi}{4}} (1 + \cos 2x)\, dx$

$\qquad = \left[\dfrac{\pi}{2}x + \dfrac{1}{2}\sin 2x\right]_0^{\frac{\pi}{4}}$

$\qquad = \dfrac{\pi}{2}\left[\dfrac{\pi}{4} + \dfrac{1}{2}\sin\dfrac{\pi}{2}\right]$

$\qquad = \dfrac{\pi}{2}\left[\dfrac{\pi}{4} + \dfrac{1}{2}\right]$

$\qquad = \dfrac{\pi}{8}[\pi + 2]$

b $\dfrac{\pi(\pi + 2)}{4(\pi - 2)}$

3 a First, $F = \dfrac{Gm_1 m_2}{d^2}$, $m_1 = m_1 m_2 = M$

On the surface, $F = mg$, so $mg = \dfrac{GmM}{R^2}$ where d = R

Then $gR^2 = GM$, so $F = \dfrac{Gm_1 m_2}{d^2} = \dfrac{GMm}{d^2} = \dfrac{mgR^2}{x^2}$

b $-F = m\ddot{x}$, then $mv\dfrac{dv}{dx} = -\dfrac{mgR^2}{x^2}$

$\int_{2U}^{0} v\,dv = -gR^2 \int_{\frac{3R}{2}}^{x} \dfrac{1}{x^2}\,dx$

Then $\left[\dfrac{1}{2}v^2\right]_{2U}^{0} = -gR^2\left[-\dfrac{1}{\frac{3Rx}{2}}\right]^{x}$

so $-2U^2 = \dfrac{gR^2}{x} \Rightarrow -\dfrac{gR^2}{\frac{3R}{2}} - 2U^2 = \dfrac{gR^2}{x} - \dfrac{2gR}{3}$

$\Rightarrow -6U^2 = \dfrac{3gR^2}{x} - 2gR \Rightarrow \dfrac{3gR^2}{x} = 2gR - 6U^2,$

or $x = \dfrac{3gR^2}{2gR - 6U^2}$

This is from the centre of the Earth. We need from the surface, so using $h = x - R$ we have

$h = \dfrac{3gR^2}{2gR - 6U^2} - R$

$= \dfrac{3gR^2 - 2gR^2 + 6U^2\,R}{2gR - 6U^2}$

$= \dfrac{gR^2 + 6U^2\,R}{2gR - 6U^2}$

$= \dfrac{(gR + 6U^2)R}{2(gR - 3U^2)}$

4 $\dfrac{5}{8}a$

5 $\sqrt{\dfrac{5g}{12a}}$

6 a Initial $KE = \dfrac{1}{2} \times 2mu^2 = mu^2$

initial $PE = -2mg \times \dfrac{3a}{2} = -3mga$

generally $KE = \dfrac{1}{2} \times 2mv^2 = mv^2$

$PE = -2mg \times \dfrac{3a}{2}\cos\theta = -3mga\cos\theta$

Using conservation of energy:

$mu^2 - 3mga = mv^2 - 3mga\cos\theta$

which becomes $v^2 = u^2 - 3ga + 3ga\cos\theta$

Force towards the centre:

$T - 2mg\cos\theta = \dfrac{2mv^2}{\frac{3a}{2}} \Rightarrow$

$T = \dfrac{4m\,v^2}{3a} + 2mg\cos\theta$

$T = \dfrac{4m}{3a}(u^2 - 3ga + 3ga\cos\theta) + 2mg\cos\theta$

$T = \dfrac{4m\,u^2}{3a} - 4mg + 6mg\cos\theta,$

at the top we need $T \geq 0$,

$\theta = 180°$

so $\dfrac{4m\,u^2}{3a} - 4mg - 6mg \geq 0 \Rightarrow \dfrac{4m\,u^2}{3a} \geq 10mgu^2 \Rightarrow \geq \dfrac{30ga}{4}$

Hence $u \geq \sqrt{\dfrac{15ga}{2}}$

b i $109.5°$,   **ii** $2.91a$

7 a $0.48\,\text{m}$

b From $F = ma$, $2g - T = 2\ddot{x}$

Then $2g - \dfrac{49}{1.2}(0.48 + x) = 2\ddot{x} \Rightarrow 2\ddot{x} = -\dfrac{245}{6}x$

So $\ddot{x} = -\dfrac{245}{12}x$ therefore SHM.

Using $T = \dfrac{2\pi}{\omega}$ where $\omega = \sqrt{\dfrac{245}{12}} = \dfrac{7}{2}\sqrt{\dfrac{5}{3}}$, we have

$T = 2\pi \times \dfrac{2}{7}\sqrt{\dfrac{3}{5}} = \dfrac{4}{7}\pi\sqrt{\dfrac{3}{5}}\,\text{s}$

c $\dfrac{1}{5}\sqrt{\dfrac{5}{3}}$ or $0.258\,\text{m}$

d $0.0880\,\text{s}$

INDEX